THE NATURE OF SCIENCE

Books by the same author

ATOMS AND ELEMENTS

NATURAL SCIENCE BOOKS IN ENGLISH 1600–1900

SOURCES FOR THE HISTORY OF SCIENCE 1660–1914

David Knight

THE NATURE OF SCIENCE

The History of Science in
Western Culture since 1600

ANDRE DEUTSCH

First published 1976 by
André Deutsch Limited
105 Great Russell Street London WC1

Printed in Great Britain by
Tonbridge Printers Limited
Tonbridge Kent

ISBN 0 233 96814 8

FOR SARAH

CONTENTS

'I . . . had ambition not only to go farther than any one had been before, but as far as it was possible for man to go.'

CAPTAIN COOK

'The most important result of a rational inquiry into nature is, therefore, to establish the unity and harmony of this stupendous mass of force and matter, to determine with impartial justice what is due to the discoveries of the past and to those of the present, and to analyze the individual parts of natural phenomena without succumbing beneath the weight of the whole. Thus, and thus alone, is it permitted to man, while mindful of the high destiny of his race, to comprehend nature, to lift the veil that shrouds her phenomena, and, as it were, submit the results of observation to the test of reason and of intellect.'

ALEXANDER VON HUMBOLDT

'The uniformity of the plan being once assumed, events which have occurred at the most distant periods in the animate and inanimate world will be acknowledged to throw light on each other, and the deficiency of our information respecting some of the most obscure parts of the present creation will be removed.'

CHARLES LYELL

I

SCIENCE AS A COMPLEX ACTIVITY

Two hundred years ago, the chemist and Unitarian minister Joseph Priestley wrote two long books on the history and present state of two of the leading sciences of his day—electricity and optics. They were intended to lead the reader, as they in fact led Priestley himself as he wrote them, from almost complete ignorance up to the very frontier of knowledge. This little book is in some ways more ambitious, and in many ways less so, than were Priestley's. It is more ambitious in that we shall be concerned not with the history and present state of a single science, but with the whole of science; and it is less ambitious in that nobody reading it will find themselves thereby equipped to do research in any science. It is not a technical work, and it has it in common with Priestley's books that no advance knowledge of current or past science is expected of the reader. As Priestley hoped to teach science through its history, so in this book we may expect to learn something of how science works, as a process and not as the impersonal enumeration of authenticated facts. Our concern will be wider than Priestley's, for we shall be looking at science as a complex human activity in order to see how it resembles, and differs from, other human activities.

We shall follow a thematic rather than a chronological approach to the history of science; trying to answer the question 'What is science?' by looking at science as a process of thinking about nature, of talking about nature, and of interrogating and using nature. That is, we shall describe science as an intellectual, a social, and a practical activity. It is only if we follow such a broad and comprehensive road that we can do justice to the complexity of science; although naturally those wishing to investigate in detail past or present science will often or usually have to restrict their view.

The intellectual and practical aspects of science have a very long history indeed, but the growth of a scientific community is a more recent phenomenon which characterises modern science, and began with the scientific societies and academies of the mid-seventeenth century—such as the Royal Society in London, and the Academy of Sciences in Paris. We cannot cover all the science since about 1660, but we can look at the intellectual, social, and practical aspects of science in the last three centuries : trying to concentrate upon episodes which seemed important to contemporaries rather than looking for heroes; and taking scientists of the past seriously, remembering that our situation and preoccupations resemble theirs in some ways, and differ in many others.

The term 'Science' means different things to different people who employ it. Down to the 1830s, when the British Association for the Advancement of Science was founded, it meant any organised body of knowledge, whether in history, theology, or chemistry. The term 'art' which we now oppose to 'science' then meant technique. Gradually since then these words have taken on their modern meaning; so that 'science' now usually includes chemistry, physics, and biology, and perhaps psychology, anthropology, and geography. Mathematics is problematic, but for most people would probably be regarded as a science.

These various disciplines have some things in common : many of them involve more or less experimental work, often done in a laboratory; and those who practice most of them would claim to be finding out about the natural world. In this sense, science is an intellectual activity to be compared to others, such as history or philosophy; it is this aspect of science which interests philosophers, is exciting to those who pursue it, and can seem dauntingly difficult to the layman.

On this view, the object of science is to explain what happens in the natural world, and to predict what will happen in the future; though exactly what it means to explain something is a question to which different philosophers give different answers. One might hope that study of science as an intellectual activity would enable us to draw formal boundaries between what is science and what is not, making this kind of study self-sufficient because one could define one's subject matter clearly and distinctly. This turns out not to be the case; not only does past science look rather different from present science—which might not matter—but there appear to be few formal resemblances between the various sciences. Applied mathematics and ecology, for example, have little in common that distinguishes them as sciences from, say, history or literature. The sciences seem to be a natural group like the natural groups of the animal and plant kingdoms, characterised by family resemblances but without clear and obvious lines of demarcation in every case.

It is for this reason that to ask—as people do—whether, for example, sociology is or might become a science seems so futile. In the past there have been within the sciences rather similar questions, such as whether the duck-billed platypus is a mammal or whether potassium is a metal. Both resembled other members of the class in many ways, but the platypus, unlike any other mammal, laid eggs; and potassium was lighter than water while all other known metals were much heavier. In the event, both were put with the class; but for some purposes it is necessary to remember that they only resemble most members of the class in a limited number of ways. The sciences are not monolithic; to make a subject 'scientific' might be held to involve bringing more rigid logic into it, or more fieldwork, or more laboratory work; to involve more application of mathematics, or the abandonment of arid and formal mathematical models; and to involve statistical, causal, or formal explanations, or to stop at classification. The range of undoubted sciences is so wide that anybody can find in them almost any rule of method; and to suppose that physics is the paradigm science, the ideal to which all the others are striving, may be characteristic of physicists but seems unwarranted for anybody else.

To a lump of potassium, it cannot matter whether we decide that it is a metal or not; and similarly the platypus will not be concerned about where we put it in our taxonomic scheme, unless perhaps mammals have more protection from hunting than birds or reptiles. That is, the question of where we put something in a system of classification really begins to matter when it makes a difference to rules or conventions. The question of whether sociology, for example, is a science has really less to do with its formal structure than with the status of its practitioners; or in an internal debate, with some of them, who are claiming the prestige accorded to scientists. In short, we can only really see the point of this kind of question if we look upon science as a social activity. Within the more established sciences too, we shall sometimes find that what looks like an intellectual debate in fact resolves itself more or less completely into a social conflict.

This may take the form of a generation gap. Darwin infuriated his opponents by suggesting that the younger men coming into biology were all turning to his theory of evolution by natural selection, and that fairly soon, as those trained in older concepts died off, everybody would be a Darwinian. Max Planck made the same point about quantum theory about 1900; and no doubt this is a general phenomenon. Generally Darwin regarded it as something out of the ordinary when the great geologist Lyell, then in his sixties, after some hesitation came out in favour of evolutionary theory. Certainly scientists do not seem to be any more keen than anybody else to drop a whole world-view upon which their life's work has been based because somebody has proposed a new and wide-ranging theory; and one reads about this as some kind of scandal only in the few cases where—as with evolution and quantum theory—the new ideas turned out to be manifestly more fertile than the old ones. Priestley was criticised by his younger contemporary Humphry Davy for embracing new hypotheses 'with an ardour almost peurile'; the majority of bright new ideas turn out wrong, and the cautious judgement of older scientists is not to be dismissed as mere old-fogeydom.

Conflicts may also arise between men of science in different countries; not merely about who discovered some phenomenon first, but also about the relationships between the sciences, and about particular theories or experimental procedures. As science, or the 'new philosophy', became prestigious at the end of the seventeenth century, so we begin to find international disputes; this was also the period of the development of nation states and vernacular languages in Europe. The *Principia* of Newton, which appeared in 1687, and set out his views of physics generally and gravity in particular, was welcomed in Britain both because it brought together into an intelligible unity the phenomena of earth and heaven, and also because it gave Britain a great man to put beside Galileo, Kepler, and Descartes. Newton's friend Halley publicised it among men of science, while Richard Bentley made its conclusions known more widely in a course of lectures against atheism, and John Locke in his *Essay on Human Understanding* provided a philosophical framework for Newtonian science.

In tolerant and Protestant Holland, with its strong links with England, Newtonian science caught on rapidly even though Christiaan Huygens, their greatest astronomer, never felt entirely happy about Newtonian physics. He had close links with France as well as England; and he believed that Newton's physics did not provide the clear and intelligible mechanical account of the phenomena that Descartes had regarded as essential. This was also the attitude taken in France; where even at the time of Newton's death in 1727 the general view was that he had provided some useful equations but no real account of what was going on. The demands made of a scientific theory in Britain and in France were different, and in the latter country more emphasis was placed upon clear and distinct mechanical analogies; and this difference was accentuated by national rivalries.

In Britain, Bentley had urged that Newtonian astronomy was particularly congenial to belief in God; for it seemed that Divine intervention had been required to form matter into stars and planets, for influenced by gravity alone it would have all collected into one vast lump; to keep up the motion in the world, which would otherwise run down; and to keep the planets stable in their orbits, for otherwise they would attract each other, and wobble in their courses. To Continental critics, and especially to Leibniz in Germany, this seemed, on the contrary, to derogate from God, who as an all-wise creator would have seen all possible contingencies and guarded against them, rather than having to keep regulating his mechanisms as they ran awry. The theological demands made of a scientific theory were thus different in different countries; and this, with the addition of straightforward national rivalry, was one of the reasons why Germans were suspicious of Newtonian theory right through the century after his death.

There was a difference, too, in the style of natural theology prevalent in different countries; Descartes deduced the inertial principle—that a body continues in a state of rest or uniform motion in a straight line unless a force acts on it—from God's changelessness, while in England Robert Boyle and John Ray began with the facts of nature and argued for God's existence and benevolence from them. This difference in style went with a difference in approach within science, Descartes particularly emphasising logical rigor while the more empirical Englishmen gave more weight to experiments as the basis of natural knowledge. Preferences of this kind have been at different times associated with different countries, and even with different universities; in the early nineteenth century the French were particularly strong in applied mathematics, which to many in Britain and Germany seemed like arid intellectual model-making; and those who built up the school of physics at Cambridge believed, in the words of one of the most eminent, G. G. Stokes, that : 'A well-established theory is not a mere aid to the memory, but it professes to make us acquainted with the real processes of nature in producing observed phenomena.' This marked a curious reversal of the eighteenth century position, where the theory of Newton—a Cambridge man—had been rejected in France because it did not seem to be 'well-established' in this sense.

The French were eventually converted to Newtonian physics by Voltaire, who in his Anglophilia poured scorn on his countrymen who still adhered to the Cartesian mechanisms, and who persuaded Mme du Châtelet to translate Newton's *Principia* into French. At the end of the eighteenth century we find the somewhat similar phenomenon, that the spread of Lavoisier's chemical theory in Germany was most rapid in the most Francophile states. In all these cases, the social explanation is not complete; for there were experimental and theoretical arguments on both sides. If we treat these situations from only one aspect, we shall leave out too much, and find ourselves only telling half the story.

To see science as a social activity involves one in more than reducing some intellectual conflicts to social ones. For if formal criteria do not separate science from other activities, one must try something else; for example, one might suggest that 'science' means what scientists do. They do experiments, and they do some mathematical reasoning in many cases; but in all cases they belong to some scientific society, read and probably also publish papers in some journal, and attend conferences or congresses to associate with others in the same field. Scientific knowledge is public knowledge, made known to a scientific community; and without such a community of people sharing many of the same assumptions and interests, the science of the last three hundred years would not have been possible. Papers are submitted to editors, who in turn send them to referees who go through them and repeat the experiments; the paper if good is then published. In universities, undergraduate and graduate students sit examinations or submit theses which must be approved by external examiners from other institutions so that standards are kept up.

These processes have caught the attention of Thomas Kuhn, who in his *Structure of Scientific Revolutions* (1970) has tried to bring together the intellectual and social aspects of science in a schematic way. Kuhn draws attention to the conservatism of the scientific community, and the dogmatism involved in scientific education, which he sees as close to drilling recruits for the army—while by contrast, arts students are more required to reason, and to display originality. We are asked by publicists to see scientists as open-minded and keen to welcome new facts, however awkward to existing theories; but if a student finds that the proportion of copper in copper sulphate is different from the figure in the back of the textbook, then his teachers will conclude that he has not found out anything interesting but has done the experiment badly.

For Kuhn, doing science is a difficult business for in any field there are numerous wrong turnings and only one strait and narrow way. Sciences emerge from collections of incoherent data when somebody hits upon a way of seeing the phenomena as a coherent whole; this vision of the new science will include some theory or hypothesis which will be a guide to research, and is by Kuhn called a paradigm. Thus he sees Franklin as the founder of electrical science because he unified a mass of previously unconnected observations about glass rods, fur, shocks, lightning, and so on with his theory that electricity was a subtle fluid that could flow around a circuit. Franklin was himself following the paradigm or example of Newton, who had demonstrated the power of theory and experiment going hand in hand in his *Opticks*; and in turn Franklin's work became a paradigm for his successors. Those taking up electricity —starting perhaps by reading Priestley's volumes—found the field circumscribed and ordered for them by Franklin, whose theory, sometimes modified to include a positive and negative fluid instead of just one kind, formed a framework within which experiments were to be planned, and observations interpreted.

Once a science has thus been set upon its way, a period of 'normal science' begins. Those learning the subject are drilled in the accepted paradigm and its associated jargon, for it would be a great waste of time if everybody had to make the mistakes of his predecessors all over again; editors and referees know what sort of reasoning and experiment is appropriate, and organisers of congresses know what are proper topics for discussion. The founder of any science must wrestle with questions which are difficult to ask, because he must frame new concepts, or take across from one field into another ideas which will bring order out of chaos. Those working in normal science do not need this kind of creative imagination, for they are concerned with questions which are difficult to answer, but can almost certainly be answered given that the researchers are persistent, well-trained, and equipped with adequate apparatus. The founder of a science is often a man who diverts attention from unanswerable questions to answerable ones; as when Galileo stopped asking why bodies should fall, and inquired instead how fast they fell.

The period of normal science may continue for a long time, but eventually it will probably happen, according to Kuhn's analysis, that unacceptable anomalies will accumulate. In no field is everything ever satisfactorily accounted for; there is always some untidiness around the edges of the most logically organised science. Thus in the nineteenth century, vestigial organs—such as the appendix in man—were an anomaly; and similarly the radiation from black bodies, and the precession of the orbit of the planet Mercury, could not be completely accounted for; but these did not seem matters of enormous importance. With dramatic new perspectives, these phenomena fell into place; the vestigial organs were an indication of evolutionary history, the radiation from black bodies was emitted in little parcels, called 'quanta', and in relativistic physics the orbit of Mercury was much more as expected than it had been in Newtonian physics.

These new perspectives constitute for Kuhn 'scientific revolutions'; and like political revolutions they mark the beginning of a new epoch. Some of the old science will survive in the period of the new paradigm, but what had seemed important will now seem peripheral, new research programmes will be begun, and very soon the preoccupations of a generation ago will seem hopelessly and incomprehensibly wrong-headed and even laughable. Some things which had not seemed problems in the earlier period become anomalous in the new. Thus for pre-Darwinians the peacock's tail had been created for the pleasure of peahens, and perhaps of squires, and indicated God's delight in beauty and diversity; Darwinians had to ask how such an encumbrance could possibly have been developed in the course of the struggle for existence, given that the peacock's ancestor must have been a bird with an ordinary kind of tail.

When a new paradigm comes in, there may be some loss as well as gain; but the new one will bring into one view more phenomena, and will lead to more experiments than the old, or it will not prevail. The opposition of older men, trained—or indeed drilled—in the old paradigm, and brought up to expect that any problem can be solved within it, is to be expected. Aristocrats rarely make revolutions, though they may—like the Duke of Orleans—support them if they think that they may thereby become king. So it is within the sciences. Those who have risen to the top by work within the old paradigm will not welcome or see any need for the new. There need be nothing in any way base about this. In the first half of the nineteenth century, Cuvier and his disciples had spent their lives in determining which species fossil bones belonged to, and had thereby brought order into geology; it was hardly to be expected therefore that such a man as Richard Owen, Cuvier's foremost disciple in Britain who had himself reconstructed the moa, an enormous extinct flightless bird, initially from a single bone, should welcome Darwin's theory in which species became unstable. Indeed those working, like Owen, in museums where they were classifying specimens endlessly, do not seem as a class to have found Darwinism particularly useful or important; they had plenty of work to do already.

In place of the view of science as a more or less steady march towards the truth or towards objectivity, Kuhn's account gives us a much more interesting picture of a series of lurches forward in a new direction, interspersed with periods of tranquil and cumulative growth called normal science. It bears a curious resemblance to theories both of history and of geology current in the early nineteenth century, and seems open to much the same objections as they were. The theory of history is Carlyle's, in which the hero plays a spectacular role and 'we petty men walk under his huge legs, and peep about to find ourselves dishonourable graves'. While this view of history has much to recommend it as against those which involve grey and irresistible forces of progress—the abortive revolutions of 1848 show that you can stop progress—it is open to grave objections in any sphere; and certainly if in the history of science we have heard of only a few great names, this seems to be because we are ignorant and not because the really important work has been done by only a handful of heroes.

The geological theory was that of Cuvier and others of his contemporaries, who invoked past catastrophes to explain apparent discontinuities in the strata. The earth's history was for them a matter of alternations of tranquillity, during which the strata accumulated, and cataclysms when everything was upset. This conception was overthrown chiefly through the efforts of Lyell, whose *Principles of Geology* began to appear in 1830. In it he drew attention to the power of the agents now at work on the earth's surface, such as waves, floods, earthquakes and volcanoes, and argued that given time these causes now active could have produced all the past changes of which we find evidence.

When we look in more detail at the history of any branch of science, we find there a succession of small changes both in data and in concepts. Franklin's particular theories in electricity were soon modified almost beyond recognition by Aepinus, Cavendish, and Coulomb; the discoveries of Galvani and Volta added enormous provinces to the science, opening physiology and chemistry to investigation by means of electricity; Oersted and Ampère brought together electricity and magnetism; and Davy and Berzelius made electric charge the basis of chemical affinity. Finally Davy's pupil Faraday killed the idea of electricity as a tenuous fluid, and saw it instead as a manifestation of force or energy. Where in all this catalogue we want to say that Franklin's paradigm was given up is a matter of taste; certainly all of Franklin's theory that survived in the 1830s and '40s, when Faraday was at work, was a general idea that electricity was an experimental science to be handled rather like any other, and that electricity might be a kind of fluid like heat was supposed to be—but this last idea was already beginning to seem old-fashioned, and there were few prepared to defend it.

Scientific revolutions are perhaps not unlike social ones; sometimes one can find a date on which some Bastille fell, but often one can only see on comparing the situation after a decade or two that some profound political, social or industrial revolution has occurred. Credit for causing it may then be given to whoever is perceptive enough to see what has been gradually happening, and to explain it to his contemporaries. Thus Helmholtz in propounding the principle of conservation of energy seems to have been stating clearly something which had been at the back of the minds of many of his contemporaries, but which they had not enunciated distinctly or seen in its full generality. Deciding what will count as a revolution, in science or elsewhere, is bound to be largely a matter of convention; it should be decided as far as possible with reference to contemporary opinion, for if people feel they have been through some kind of revolution then that is an historical fact to be recorded and accounted for—whether we believe they have been through one or not.

If on the contrary we impose our criteria of revolutions upon the past, then we shall find ourselves arguing about whether or not some episode was a real revolution or not; and this leads us into an introspective examination of the schema and away from the attempt to see past events as contemporaries saw them, and thus really bring sense out of them. Kuhn's schema is a very valuable one because it gets us away from the idea of science as steady 'positive' progress opposed only by reactionaries, theologians, or metaphysicians, or as perpetual revolution in which every step involves extraordinary creative energy; and in our terms, because it synthesises the intellectual and social aspects of science. Armed with the idea of scientific revolutions and of normal science, one can the more easily see science as an expression of culture at a particular time and place, and not as a better or worse approximation to some ideal and impersonal truth.

Any schema is going to let us down if we take it more seriously than it deserves, because the history of science is—like any other kind of history—the story of particular people doing particular things in particular situations. Except in the most general way, we should not expect to find any pattern of development in science or in anything else; no two situations are ever quite alike, and the historian should merely hope that the sort of approach that has cast light upon one past situation may do it again. This kind of analogical reasoning is after all used by scientists when they take a paradigm from an established field and see if something like it will work in a new one. Similarly, Kuhn's theory makes us look at some part of the history of science, and ask whether we are encountering there the relatively predictable and cumulative advance associated with his normal science, or whether it fits better the idea that some conceptual revolution is in progress, and for the moment all coherence gone. Both his normal science and his revolutions brought about by great men are caricatures; they must not be applied in some procrustean manner as though they were a kind of blueprint to which the past must be made to conform, for that would lead to a new scholasticism.

In a revolution the baby is often let go with the bath water, and in applying Kuhn's theory of revolutions to any situation there is loss as well as gain. This is familiar in other contexts, such as revolutions in the visual arts; if geometrical perspective is demanded in pictures, then one will get more convincing landscapes and townscapes, but less impressive pictures of Christ in Majesty. Kuhn's schema is very helpful in analysing episodes in the history of science which are either very dramatic or extremely dull, but rather less useful for those longer periods in the history of science when conceptual change and experimental discovery have gone on together so as to produce a vast change over a number of decades.

Kuhn's schema brought together the intellectual and the social aspects of science; but within it the practical side of the sciences are liable to be neglected. To get a rounded view of any phenomenon, we must look at it from different sides, and Kuhn's perspective must be complemented with as many others as we can find. An alternative schema to Kuhn's which has been found very widely useful is the Marxist one; according to which technology, or the means of production, is fundamental, and the intellectual and social aspects of science form a superstructure. This interpretation is in line with the view many people take of science; which they would see as a utilitarian or practical activity, which was widely expected from the days of Francis Bacon to be essentially benevolent but is now often seen as threatening.

If we look at the writings of many scientists of the past and the present—and even more if we look at propagandists for science—we find this stress upon utility there. Support comes to science from governments and from industry because of its supposed usefulness, in the long run if not in the short. Particularly in the last hundred years, industries have grown up which are based upon recent science; but even earlier this sometimes happened, as for example in the early nineteenth century when the gas industry began within a generation of the isolation and recognition of different gases—they had previously been thought of, if at all, as good or bad samples of air.

Often industry has posed problems for both applied and pure scientists, and because it can provide the money to solve them it has ensured that these were the problems which did get investigated. The most important concern of the astronomers of the eighteenth century was the fixing of longitude so that ships could, by simple observations, find out where they were at sea; the pursuit of this objective in the great observatories, and on expeditions sent out to determine the exact form of the earth, led to various discoveries in optics, magnetism and geology as by-products of the ultimately successful quest. Later, it was William Smith, a canal engineer, who made the first geological map of Britain, following the strata by means of the fossils contained within them; while problems connected with bridges, steam engines, and telegraph cables provided work for physicists, chemists, and oceanographers as well as engineers.

One can well argue that the distinction between pure and applied science is a relatively new one; that down to the middle of the last century, there was little idea that there should be any division of labour between the pure scientist who discovered something, and the applied scientist who adapted the discovery to some useful purpose. Berthollet, the great French chemist, worked on chlorine and applied the results to the bleaching industry; Davy adapted new discoveries about flames in inventing his safety lamp for coal miners; and in America Joseph Henry perfected the electromagnet and designed one for use in separating ores. Men of science took it for granted that any scientific discovery would have its use in time, even if at present no use could be foreseen.

There were also, as there still are, very close links between medicine—an applied science—and the various biological sciences; where medical advances have led to increases in biological knowledge, and vice versa. In the nineteenth century, an example was the close link between microbiology and the study of epidemics; a link that led in the latter part of the century to the germ theory of disease. It was only with the specialisation, which has been such a feature of science since the middle of the nineteenth century, that just as natural philosophers separated into chemists, physicists, physiologists, and so on, some turned into applied scientists.

Science is a practical activity not only insofar as its results may be turned to use, but also because it depends for the most part upon observations and experiments. These need apparatus —the germ theory, for example, could not have been worked out until powerful microscopes with which the bacilli could be identified were available. In the same way, the determination of longitude became possible with the invention of Hadley's quadrant which made accurate celestial observations possible at sea; and then of the marine chronometer which kept good time on a long voyage through various climates. Less directly, the coming of gas and then of electric lighting made laboratory work much more convenient; while the Bunsen burner was one of the great innovations in the history of science, enabling the chemist to do without furnaces and spirit lamps, and to get heat just where it was wanted—and incidentally making spectroscopy practicable as a method of analysis, because the sample could readily be heated to incandescence.

We do not need to be vulgar Marxists, supposing that theory in science is a mere excrescence upon technology; but we can hardly deny that theory and practice are interdependent in the sciences. The Marxist schema again would lead us to expect that it would be maritime powers that would be most interested in means of finding longitude, and hence would be supporters of astronomy; and would remind us that microscopy could only flourish when there was an adequate technology for microscopes to be made. That is, we are reminded that science is subject to financial and technical stimuli and constraints; and it may also be drawn into social conflicts. Darwinian theory was in a sense an application of *laissez-faire* doctrine to the realm of nature; since it worked there, it provided a justification, under the guise of 'social Darwinism', for ruthless social policies. Just as Kuhn's schema can help us to look beneath intellectual debates for social tensions within the scientific community, so Marx's can help us to see science reflecting and contributing to tensions in society in general.

The concept of an ice age came late into geology, and to see why we have to invoke all the three aspects of science. It is an unsurprising social point that it should have been a Swiss geologist, Louis Agassiz, familiar with existing glaciers, who recognised in the erratic boulders and polished rocks all over northern Europe and North America evidence of past glaciers; and published his conclusions in his *Études sur les Glaciers* in 1840. The theory aroused at first a certain amount of opposition; partly on intellectual grounds, because what Agassiz recognised as glacial detritus had been previously described as 'diluvium' and put down to the effects of Noah's Flood. By 1840, Lyell in Britain had persuaded most of his contemporaries to leave the Flood out of geology, and to explain past changes in terms of causes operating at the present day. But, for social and economic reasons, geologists in Britain were unfamiliar with Alpine glaciers though they knew a good deal about Polar ice. This was because since the time of Captain Cook there had been a series of expeditions to the Polar regions, looking for a North West passage and doing geophysical surveys; and because of the economic importance of whaling, which supplied the oil needed for lighting. Thus much was known and published about ice, by explorers and by a whaling captain who became a Fellow of the Royal Society, William Scoresby, in his *Arctic Regions* of 1820. Lyell therefore invoked icebergs and ice rafts as the agents which had carried erratic blocks around, because this was a cause with which he was familiar from the literature

25

in English. Once Agassiz had published, and made those in Britain and North America look to glaciers in Switzerland as a geological agency previously neglected, his idea of an ice age gained rapid acceptance because it fitted well with Lyell's approach to geology.

Those studying the history of science and technology have often sought for and, to their satisfaction, found evidence of progress. This idea goes back to the seventeenth century: when it seemed that in literature, architecture, and government the Moderns were at best equal to the Ancients, and were usually inferior to them; whereas in the knowledge of nature it could reasonably be argued that the work of Harvey, Galileo and their contemporaries was a clear advance over what had been done in antiquity. By then it was clear too that there were better machines available than there had been; horses were more efficiently used, iron was much more generally available, guns had come in for warfare, and wind- and water-mills ground corn. The invention of the telescope and the microscope put new instruments in the hands of natural philosophers, and enabled them to cross a technical frontier which had held up their predecessors. To establish that the Milky Way was a cluster of stars, that other planets—notably Jupiter—had moons as well as the earth, and that gnats hatched from eggs rather than slime, optical devices were necessary. The invention of new instruments that enable us to answer questions that our ancestors could not answer has continued, and does undoubtedly constitute a kind of progress.

By the eighteenth century, it seemed that the theories of scientists were much more powerful than those of their predecessors had been, that their apparatus was better, that scientific academies and societies had been perfected to carry forward the work, and that technology could promise endless benefits. Science would overcome superstition and ignorance; its societies were international and enlightened; and its associated technology would produce abundance. These conclusions were not ill-founded. Newtonian physics was far better at prediction and explanation than Aristotelian science had been, so that not only eclipses but also the return of comets could be accurately predicted, and accurately observed through telescopes. European capitals gradually acquired academies of sciences, each of which published its *Transactions* and gave lustre to the court. The steam engine, new agricultural implements, and then textile machinery did indeed transform society by the end of the century. Scientific knowledge seemed to be strikingly cumulative, certain, open, and useful.

In the nineteenth century therefore the idea of progress came from the sciences into history generally; and many of the greatest historians—notably Macaulay—wrote history as a record of progress down to the present. There had been some alarming moments—not everybody deserted James II immediately and rallied to William of Orange—but upon the whole history showed progress steadily leading up to the mid-nineteenth century, and doubtlessly continuing into the future. This kind of history is now called Whig history; and most historians in our century have turned against it. They have found it more illuminating to look at the phenomena described in the sources as far as possible through the eyes of contemporaries; rather than seeing them as positive or negative steps towards the present state of affairs.

Magna Carta becomes an episode in a power struggle rather than simply the foundation upon which the liberal legislation of the nineteenth century was built. In looking at any source, one asks first what was its context; why was this book written, this building put up, or this charter written or forged? Similarly, in reading the real or alleged utterances of some statesman, one does not simply judge the sentiments as admirable or deplorable by our standards, but one asks what interests he represented, what his audience was, and what was the precise point at issue?

The same process is happening in the history of science, though there are recent works which set out to describe at some length, and often tediously, the origin and development of concepts now important in the sciences. For those engaged in science, such exercises can be useful; the review of developments in the last twenty years or so, for example, is a feature of some scientific journals, and can be used to give a sense of direction to those engaged in some field where there has been rapid change. Over a longer period, there can be nothing but good coming from chemists learning about the history of atomic theory or of thermodynamics, for example; and naturally they may find it easier to begin from where they stand and go backwards rather than to begin with an attempt to understand the positions of the natural philosophers of the early nineteenth century. Their history will then be a matter of 'contributions' and 'discoveries', of mistakes, hold-ups, and blind alleys, and of truth in the end being mighty and prevailing; which can be an exciting story, but must seem sometimes remote from what they do themselves. Such history may encourage the view that we are dwarfs on the shoulders of giants; that there were in the past doughty warriors who fought terrible battles against all sorts of demons, and—rather sadly—made the world safe for us to live in; or else that past men of science were really rather foolish creatures, who could not see the truth although it lay open before them, but at whose quaint ideas one must do one's best to keep a straight face.

An example of where belief in progress in science can lead is the remark made by C. P. Snow in an essay in the *Times Literary Supplement* that soon any schoolboy will know more science than Newton did. There is a sense, clearly, in which this is true; for Newton knew nothing of motor-bikes, electric torches, or atomic energy, and such elementary physics as Ohm's Law was not published until a century after his death. But in the same way, one might argue that Gladstone or Lincoln knew less about politics than do any Member of Parliament or Congressman today, because they never learned how to put themselves across on television. It is only if we think of science as an accumulation of 'information' that we could think a schoolboy's knowledge of it could exceed that of a Newton. A recruit's knowledge of arms drill with the latest kind of rifle would soon exceed Napoleon's, but one would not want to say that he knew more about war.

In order therefore to take Newton seriously—and the same applies to any past man of science, even if not especially distinguished—we must take science more seriously, and see it as a more or less comprehensive world view. The schoolboy is crammed with information, but has not yet had time to do more than absorb the received opinions about how this is all to be fitted together. In contrast, the mature Newton read deeply in theology, prophecy, alchemy, and economics as well as physics, and developed a coherent picture embracing these various elements of his experience. His world was one in which God was omnipresent and active; and this insight provided the key which revealed the interconnectedness of things. Newton's world view was, like that of anybody who thinks at all, idiosyncratic; to his eighteenth-century successors it seemed on the one hand heretical, because he was a Unitarian and a pantheist, and on the other curiously metaphysical and theological rather than modern and mechanical. To older contemporaries, the mixture of elements in Newton's thought would have seemed unsurprising, though his particular synthesis remained his alone; natural magic was a prominent feature of the intellectual life of the seventeenth century, while theology was still Queen of the Sciences, and since archaeology was not yet capable of dating past occurrences, Newton's attempt to do so using the Bible, Greek stories such as that of the Argonauts, and astronomical data was not unreasonable. In later life, as Master of the Mint, he was in charge of the recoinage; and ruminated over such problems as whether it might be possible to make alchemical gold purer than the ordinary metal.

Down to the middle of the nineteenth century, scientists were called 'natural philosophers'. The change of name was not simply a replacement of one term by another with the same meaning, but was something that went with specialisation. The natural philosopher was expected to develop and present a world view; his object was not merely to acquire information about some part of nature, but also to acquire wisdom. That we have lost a coherent picture of the world seems indubitable; and we have also lost, in the severence of science from theology, a moral direction in science. Specialisation involves narrowing, and this is the reverse of the coin the obverse of which is scientific progress. We live in a world in which one doesn't get something for nothing, and every advance brings its attendant disadvantages. The mechanical philosophy of the seventeenth century brought advantages in the study of the physical sciences, but meant that in the postulation of *l'homme machine* older physiological and psychological insights were lost to view; even within the sciences, progress means loss as well as gain.

If, then, the sciences are more than the accumulation of data, and represent an attempt to attain a comprehensive picture of the world, we may be reassured that science will never be complete. The British government provided funds for the Geological Survey at first in the belief that this was an operation that could be carried out once and for all in a few years; but even such descriptive science turned out to be open-ended, because the interesting things one is looking for change with time —different theoretical and economic situations necessitate shifts in outlook. The situation is a more general one, for in science and technology—as in life generally—there are no final solutions. The sciences are an endless process, for every step poses new questions.

Thus Copernicus' theory, in which the sun was put at the centre of the system of the heavens, with the earth and other planets going around it, solved the problem of why the planets seem to move irregularly across the sky, sometimes stopping and moving backwards for a time. These 'stations and retrogradations' were to be expected if the earth and planets were encircling the sun, as the different planets in their various orbits lap the earth or are lapped by her. If the earth was in motion, though, gravity became a problem—one might expect that unsupported bodies would fall into the sun; and as the earth was supposed to be rotating on its axis every day one might expect that bullets would travel much further towards the east than towards the west. Copernicus' theory in short demanded a whole new physics; and so it has been with discoveries since. With every advance, things that had been mysterious or only accounted for by *ad hoc* hypotheses are fitted clearly into the scheme of things; while other phenomena, which had not been problematic, need a new explanation. In technology, we find the same phenomenon; the solution to any problem brings new problems in its train. All this seems a matter for rejoicing rather than repining, for building an edifice that was soon going to be complete might be rather dull.

The old certainties within science itself have been gradually lost since the seventeenth century. There used to be much emphasis placed upon the discovery of laws of nature; which were the analogue in the natural world of the natural law which should regulate society. Francis Bacon's writings on the philosophy of science seem to have become influential in the 1640s at the time of the English Civil War, just as his view that the King was subject to the law and not above it gained ground. There was a natural law which kings must obey, and laws of nature which governed matter. Just as kings must command their subjects only in accordance with law, so in a famous aphorism Bacon urged that men of science must 'command nature by obeying her laws'.

In the sciences we do indeed meet 'laws of nature', but they are a rather curious collection. Avogadro's Law states a proposition about molecules, which are hypothetical entities, and should probably be described as an hypothesis instead; other laws seem to be neat empirical generalisations, and to say 'It's a law of nature' is a way of closing a discussion. It is chiefly in elementary science that we meet laws; the most famous being Boyle's Law and Ohm's Law. Boyle's Law states that the volume of a mass of gas at constant temperature varies inversely as the pressure—the harder you squeeze it the smaller it gets; and Ohm's Law that the electric current flowing through a given resistance varies with the potential difference applied—if you put two volts across a coil of wire, you get twice as much current as if you put one.

Both these laws were arrived at as empirical generalisations, though no doubt their authors had in mind some such relationship when they began their experiments. The laws are still taught as such, and in elementary physics classes children still verify them by experiment. Since in any experiment or measurement there is an element of error, the laws are more accurate than the individual experiments upon which they are based. But when we widen the range of phenomena upon which the law was based, we find that it breaks down; some philosophers of science see scientists as attempting to falsify hypotheses, theories, or laws of nature in their researches, but really what seems to happen is that one experiments to find the range over which a law holds. Thus Boyle's law holds for gases only at relatively low pressures and high temperatures; and Ohm's law does not apply to liquids through which electricity passes by electrolysis rather than by conduction. A gas which obeyed Boyle's law exactly would be an ideal gas; calculations for such an entity are much easier than for real gases which behave in a less tidy manner, and the results can be adjusted to fit real cases. The laws circumscribe what can be done—no gas expands as the pressure is increased—but they do not, like the fiat of a powerful despot, lay down what must happen in all cases.

This is not unlike our laws. Henry Maine, a contemporary of Darwin, wrote an influential study on ancient law; essentially concerned with the problem of how, since a Roman was forbidden to alienate his patrimony, land and houses changed hands in Rome as easily and frequently as elsewhere since then. He found that Roman lawyers had evolved ways around the problem; the law here was a kind of convention, which meant that if one wanted some end one had to go about it in a special way, and not some system of absolute regulations. The relationship of laws of nature to technology seems to be similar; there are a few things one cannot do, like getting energy out of perpetual motion machines, but for the most part laws provide a framework within which one must be circumspect and not a list of things allowed and prohibited unconditionally.

In the sciences, therefore, one confronts a situation in which more or less appropriate conceptions are imposed upon more or less accurately determined facts, rather than a collection of indubitable truths. The march of science is a process that will never end, rather than the occupation of a promised land that will sooner or later be completely subjected. Science is as much an expression of culture at a time and place as a collection of true statements about the world; and this will affect the way we write or understand its history. Those who have looked for scientific progress have tended to follow the development of concepts or of sciences through time; tracing for example the atomic theory, or the science of chemistry, perhaps from the Babylonians up to the present day or perhaps for a shorter period. This approach involves the assumption that there is a fair amount in common between the science at the beginning and end of the period, which may or may not be plausible; and it leads one to look for the first discoverer of some substance or effect, and for 'anticipations' of ideas currently in vogue, which can be rather pointless and barren ways of proceeding.

Thus it does not very much matter whether Darwin was the first person to propose an evolutionary theory, or Dalton the first to propose an atomic theory—in fact, neither of them were; what is important is that they got their contemporaries to take these ideas seriously as a part of science. This calls out in part for an intellectual explanation, for the evidence available to them was more copious than it had been to earlier thinkers; in part for a social one, because for one reason or another these were people whose hypotheses had to be taken seriously and who could not be ignored; and partly practical, for on the one hand apparatus was becoming available for quantitative chemistry as it had not been in the days of Boyle or Newton, and on the other the agricultural revolution of the late eighteenth and early nineteenth centuries had yielded much knowledge of stockbreeding, and the scientific expeditions made possible by the conquest of scurvy had brought much new knowledge about the distribution of plants and animals.

It may often thus be rewarding to look across a range of sciences at one time, rather than at one science through a long time; for at one point of time many of the factors will be constant, and we can hope to get a fuller understanding of what science seemed to be to its practitioners. Interactions between the various sciences have always been interesting, for the status of the different sciences has varied and at different times different sciences have been the most exciting or most popular. Particularly before the increasing specialisation of science became so obvious in the later nineteenth century, men of science worked in what now seem to us an extraordinarily wide range of sciences; to understand their work it is best to look across the whole range of science in their day, for if we look at the history of one science we shall get a curiously restricted and distorted view. There is much to be said for examining carefully one of the strata of history rather than trying to drill a narrow hole down through all of them. If one does this, one becomes more aware of the richness and variety of the sources available, the need to try different viewpoints to get a clear perspective, and the uniqueness of any historical situation despite its resemblances to and relationships with others; and one is not tempted to take too seriously any historical schema, such as that of Kuhn or Marx, which seems to lay down general laws of historical development.

We shall in this book be concerned with the past and present state of science, seen under its three aspects as an intellectual, a social, and a practical activity; the six following chapters giving different perspectives on the history of modern science, and the Epilogue bringing the story up to our day and bringing the different threads together. Throughout, we shall be concerned with the natural sciences—that is, the physical sciences, the life sciences, and the earth sciences—rather than with the social sciences, the history of which is distinct from that of the natural sciences and which do not seem to constitute a unit with them whatever our viewpoint. On the other hand, science, with its various aspects, is so wide ranging an activity that almost everything in recent history has some connection with it, and we must not be in a hurry to draw boundaries around it which would prevent our understanding of what happened in the past, or happens in the present. We shall try to concentrate in this essay upon the norm, on what usually happened or happens, rather than simply upon exceptional men and happenings.

2

SCIENCE AS EXPLANATION

Curiosity was and is the chief impetus to science. The world is a booming, buzzing confusion, in which some order must be discovered or upon which some order must be imposed. Attempts to discover or impose order which go beyond the immediate needs of ordinary life constitute science; and various rules of so-called scientific method have been drawn up to guide those engaged on such attempts. When objects or happenings have been ordered, then the way is open for predicting what will happen; then we know for example, like Alice, that 'if you drink much from a bottle marked "poison", it is almost certain to disagree with you, sooner or later.' We can make predictions even when we know rather little about what is going on; Alice knew little about the physiological action of various poisons; and the early weather forecasters, too, had little detailed knowledge of meteorology to guide them. While making predictions is useful and encouraging (at least if they turn out right), it is not the chief end of science; for to satisfy curiosity, we want an understanding of the phenomena, or appearances. That is, we demand an explanation; and it is with explanation that we shall be in this chapter concerned.

We are sometimes apt to suppose that there is one kind of explanation which is properly called 'scientific', and that any other kind is unscientific. But if we look at the history of science, or at current science, we find that there have been, and are, various kinds of explanation in vogue at various times and in various sciences. Science is an ongoing process, and different kinds of explanation are appropriate in different situations; and scientists are like everybody else influenced by fashion, so that in the years just after Newton published his work on gravity such mathematical and mechanical explanations became the norm, while Darwin made evolutionary explanations popular in fields remote from biology. The various kinds of explanations which have been used, and many of which still are used, in the sciences can be classified; and it is worth doing so, even though we find that there are very few actual explanations of real phenomena which fall precisely into one of the categories. Science, like nature, does not quite fit our tidy patterns; and naturally the most creative men of science are those who have not simply followed the rules laid down by their predecessors.

The scientists of the modern period inherited various views about explanation, which came ultimately from the Greeks. These ideas were explicit in the writings of philosophers, and implicit in those of scientists and doctors, which were translated into Latin from the twelfth century on, and printed in good texts in the Renaissance of the fifteenth and sixteenth centuries. The dominant traditions were those stemming from Plato and from Aristotle; but the Renaissance was a time when the greatest men sought a unified world view, and the Platonic and Aristotelian traditions cannot be completely separated. Then as later, it was the man in touch with more than one tradition who was most likely to be disturbed by weaknesses or anomalies in one of them, and to go on and make discoveries.

In general, the Platonic tradition placed more emphasis upon mathematics and deduction; while the Aristotelians were insistent upon the collection and subsequent ordering of facts. They put their emphasis upon logic rather than mathematics. By the early seventeenth century, the scientific side of Aristotle's work was relatively neglected in most universities, and was taught as dogma rather than as empirical generalisations. Though this seems not to have been the case at some Italian universities, most notably at Padua; where William Harvey went to study medicine in the opening years of the seventeenth century.

Aristotle's biological writings remain a model. He performed numerous dissections, particularly of fish, and gave careful descriptions of many creatures; paying particular attention to the way they brought forth their young. He divided creatures into those whose young were born alive, those which laid eggs from which miniature adults hatched, those whose eggs produced a grub which was transformed into the adult form later, and those whose generation was equivocal, which might be spontaneously generated from mud. He recognised the need to use a range of criteria in classifying animals; and thus separated dolphins from fishes, and bats from birds, despite their apparent similarities. He distinguished homologous organs, which have the same structure though they may differ in function, from analogous ones, which have the same function but differ in structure. Thus hands, whales' flippers, and bats' wings are homologous; while bats' wings and bees' wings are analogous.

Aristotle was thus able to provide the basis for a natural classification of animals, rather than an artificial one such as that which puts deer, geese, and salmon together because they are all good to eat, or groups eagles, butterflies, and flying fish because they all fly. But in his biological writings he emphasised the functions of the various parts of the animals. He was not content with simply describing an organ, that is, but went on to give an account, or an explanation, of the role it played in the life of the creature. Aristotle firmly believed that Nature does Nothing in Vain; that is, that there is in nature nothing pointless, or that has simply come there by chance. All the parts of an animal must therefore have a function; and it is the task of the zoologist to find it out. This is the kind of assumption that cannot be disproved, for if we cannot find a function for some part the fault may be in us; it cannot be proved either, but Aristotle believed that there were so many cases where the parts of animals were so well adjusted that it would be folly to deny that Nature does Nothing in Vain.

Aristotle therefore believed in teleological explanation; that is, he thought that something was only explained when the purpose, end, or point of its existence was made clear. To describe the shapes and positions of the parts of an animal was useful, but to explain them one must find what function they perform in the whole life of the creature. Analysis is thus not enough, for nothing can be explained in isolation. The end or purpose of anything is called its final cause. In Aristotle himself, these are not man-centred; the final cause of the birth of a calf is that an adult animal should come into being, and not that we should have milk or beef. His medieval successors were less detached, and in their bestiaries the final causes of animals are to feed or clothe us, or to teach us moral lessons; the butterfly, for example, which starts life as a miserable crawling thing, dies into a crysalis-coffin, and rises as a beautiful winged creature, is an allegory of our life, death, and resurrection.

Aristotle's teleological beliefs were shared by the great physician Galen, whose voluminous writings formed the basis of most medical courses in the Renaissance. He investigated, for example, the connection between the kidneys and the bladder; showing how the urine flows down the ureters, which are so arranged that it cannot flow back again. Galen was less happy in accounting for the relationship of the heart, lungs, liver, and blood vessels; and at Padua from the 1540s a series of anatomists beginning with Vesalius began to follow Galen's methods rather than his text and to elucidate this problem. They were guided by the same belief in final causes, and in Nature doing Nothing in Vain, as Aristotle and Galen had been.

William Harvey, who finally solved this problem after his return to England and published his book on the circulation of the blood in 1628, is—and was in his own day—reckoned one of the great men in the rise of modern science. It is surprising therefore to find that he was an admirer of Aristotle, whom we often find in history books as the symbol of arid scholasticism against which science had to fight. Harvey employed various arguments to justify his view that the blood circulated; but those which seem to have carried most weight with him and with his readers were the teleological ones. If we suppose that the blood circulates, then all the valves in the heart and the veins have a clear function which they perform efficiently—namely, to keep the blood from flowing back. Similarly, the veins and arteries differ in structure; if the blood circulates, then the arteries have to sustain a higher pressure, and so that is why they are stiffer.

Teleological arguments can lead to silliness, as Moliere showed in making the doctor say that opium sent us to sleep because it had a *virtus dormitiva*. But as used by Galen and Harvey, to relate organs by taking function into account, they are not open to this objection; and indeed are today widely used in biology, although those philosophers of science who think that physics is the only real science are unhappy with them. Teleological arguments can also be pushed a stage further; for one can argue that the order which we see in the parts of plants and animals must be the result of design. As William Paley put it about 1800, if we came upon a watch lying on the ground, and saw how all the parts were adjusted to work together, we would not suppose that the watch was the outcome of chance collisions of atoms; on the contrary, we would say that somebody had designed it and made it. Since living creatures are every bit as complicated and well-adjusted as watches, Paley concluded that they must be the work of a wise and benevolent creator.

Darwin's *Origin of Species* of 1859 brought about a new attitude to teleological explanation. On the one hand, it encouraged the use of functional analyses of plants and animals; while on the other it seemed to provide a way round the argument from design. Because of the pressure of natural selection in the Darwinian world, animals whose parts were not well adjusted to one another would perish, and only organs useful even in their most primitive condition to a creature would be developed. It will therefore follow that in general, as Aristotle and Galen had urged, all the parts of a plant or animal will have a function. Darwin himself, following this doctrine, was able to elucidate the sexual parts of orchids which ensure cross- rather than self-fertilisation; and to explain the structures of some insectivorous plants, for he refused to believe that they would have developed pitchers unless this had been of advantage to the plant. Darwin's disciples called this the new teleology, and rejoiced that one could now use functional explanations without being mysterious. But Darwin's teleology was superior to Aristotle's because he could account for some few organs for which no use could be found, such as the appendix in man, which would otherwise seem a poor feature of the design of the organism. These, like the homologies which Aristotle had noticed, were evidence of evolutionary history, and were the remains of organs which had been useful in some remote ancestor. Darwin's theory could thus account for most organs being useful, and for a few being useless.

Biologists could then continue—or begin again—to use teleological explanations, conscious that these could be rephrased in terms of the struggle for existence if one were pressed by some enemy of final causes. In Darwin's own day and since, there has been no complete agreement as to whether the development of higher organisms from lower ones requires some designer. Darwin himself invoked the analogy of a stockbreeder improving his cattle by selection of the best animals to breed from; but he did not himself posit a cosmic stockbreeder, and his evolutionary process is a blind one leading simply to creatures better adjusted to their environment. The tapeworm is as much the product of evolutionary forces as the man it inhabits.

In Darwin's circle, the agnostic T. H. Huxley was prepared to accept this kind of world; and like Darwin denied that this involved belief in mere chance as the basis of evolution. For Darwin, the choice was not simply the traditional one between chance or design; he urged that there was a third alternative, development in accordance with law. His American friend, the botanist Asa Gray, could not believe in laws without a lawgiver; and an increasing number of theologians as the century went on came to believe that the conception of God as the initiator of an evolutionary process was grander and juster than to see Him as the designer of every particular species of mollusc. But the lawgiver was not strictly necessary to the system; and Darwin's contemporaries and successors were faced with the nightmare of a world without apparent moral order.

Despite Harvey's use of teleological reasoning, the use of final causes in scientific explanation greatly decreased in the seventeenth century. This was not because men of science then did not believe in God; most were religious, and did their science in part as a religious duty of revealing divine benevolence. But to Harvey's patient, Francis Bacon, and other critics of Aristotelian science final causes seemed like barren virgins. Enquiry was stopped by some teleological explanation when it ought to have been pushed further. This pushing further was expected to lead to a mechanical explanation; or at least to a mathematical account of the phenomena. Associated with this mechanical philosophy was a new notion of 'cause', which for Aristotle had included the final cause, the material cause (that is, the matter must be present), the formal cause (or plan), and the efficient cause (or circumstance which sets the process going). None of these correspond to our ordinary or technical use of 'cause' as the sufficient and necessary condition of the effect—a usage which goes back to Bacon's contemporary, Galileo. Instead of a complex of conditions (appropriate to the biological sciences), the new philosophers sought single causes and thought of causal chains stretching backwards through time.

Mechanical explanations were closely associated with the atomic theory; which at the beginning of the seventeenth century seemed heretical, but by the end of it was accepted in some form by most men of science. The source of modern atomism was Lucretius' poem *de Rerum Natura*; but whereas the atomists of antiquity had been content to show that the various properties of things could be accounted for by supposing them to be made up of particles having simpler properties, their seventeenth-century successors tried to apply the theory in a more definite way. Ancient atomists were happy if they could think of two or more plausible explanations; the moderns rejected a science of likely stories, and believed that they could find the single true explanation of a phenomenon. The rise of modern science in the seventeenth century can be seen as a pursuit of certainty.

Galileo used ideas derived from atomism to explain the effects of heat; he declared that motion was the cause of heat, and envisaged little particles as responsible. Descartes in the 1620s and '30s described a mechanical world, in which the motions of particles lay behind all the appearances. Unlike the atomists, Descartes postulated a full world; theirs was mostly void space in which the atoms moved, while his had no empty space at all, the gaps between the particles being filled with subtle matter. What seemed to be action at a distance, such as the sun's power to hold the earth in its orbit, became intelligible for Descartes' disciples; who saw the sun (as it rotated) swirling the subtle matter around in a vortex, which carried the earth and the other planets around.

This might seem to us just another more or less likely story; but Descartes hoped to achieve certainty by his famous method of systematic doubt. He found that he could not doubt his own existence as a thinking being; as he declared in his famous aphorism, '*Cogito ergo sum*', I think therefore I exist. He next proved that God must exist; both because our imperfect minds could not have the notion of perfection unless a perfect being had put it there, and also because God's existence seemed necessary in a way that the existence of other things is not. This is the famous ontological proof of God's existence. We know that material objects come into being and decay, and that living creatures are born and eventually die. There is no absurdity in saying of any object that there was a time when it did not exist, and that there will be a time when it no longer exists. But it would be absurd to say that of God, understood as the creator and sustainer of the entire universe. To Descartes, then, it seemed that God must exist; and that any really clear and distinct idea must have been planted in his mind by God, who would not deceive him. Whereas teleological thinkers saw order in the world and attributed it to God, Descartes began with God as the guarantor of order.

Unfortunately the physics of Descartes was not as certain as its author hoped it would be. In a famous passage, he derived the law of inertia—that a body continues at rest or in uniform motion in a straight line unless a force acts on it—from the unchangeability of God. But when he came to do optics, which was in many ways the best part of his physics, he began with the clear statement that light must travel with infinite velocity; and then in accounting for refraction, wrote that light must go faster in water than in air. He illustrated his account of reflection and refraction with little pictures of men playing tennis; his hypothetical light particles behaved in some ways like little tennis balls. Rather than achieving certainty, he was using models which accounted for some phenomena but not for all.

The fit of Descartes' models with the world could not be tested in his own day; but a generation later Roemer demonstrated that light travels with a very high but not infinite velocity, and in the nineteenth century Foucault showed that light travels faster in air than in water. The atomic explanations proposed by Boyle and Hooke, in the generation after Descartes, for the facts of chemistry, biochemistry, and mineralogy were also untestable before the nineteenth century. The mechanical, atomistic philosophers of the seventeenth century were able to paint a world picture which seemed convincing to most of their contemporaries, at any rate by the end of the century; but their atomism was a very general doctrine, which could explain very few phenomena in any detail. In Germany, Leibniz rejected strict atomism in favour of a theory of monads; ultimate atoms or corpuscles were supposed to be all alike, or of very few different kinds, whereas the monads were all different from one another. But although natural philosophers might differ about the existence of void space and the number of kinds of particles, they agreed upon the whole that the phenomena of nature were to be explained in terms of the arrangements and motions of particles.

Descartes made his reputation when he discovered that propositions in algebra could be interpreted geometrically, and vice versa. Before his time, algebra had seemed a useful but not rigorous branch of mathematics, and geometry the branch where satisfactory proofs were to be found. Now algebraic propositions could be shown to have the same certainty as geometrical ones; and a new world of mathematical physics was opened up. In the Aristotelian tradition, mathematics played a subsidiary role in science. It was the philosopher who worked out the physical explanation, while the mathematician could at best devise a method of saving the appearances, or fitting the facts. On this view, his equations or constructions would have no claim to physical reality; they would be judged simply by their simplicity and convenience.

This was the view taken by conservatives faced with Copernicus' new astronomy, which was published in 1543; indeed the publisher had attached a preface to the book urging that Copernicus was not concerned with physical reality, but had suggested that the earth moved around the sun simply to make astronomical calculations easier. This seems to have been the view of astronomy taken by Ptolemy, who had in the second century AD proposed the system of epicycles based upon a central earth which was still used by Copernicus' contemporaries; he did not really believe in a 'big wheel' universe as a physical reality, but did show that epicycle constructions enabled planetary positions to be predicted. Copernicus, on the other hand, did believe that the sun really was in the centre of the system; and the preface to his book did not do him justice. He did however continue to use epicycles; and it was not until in the early seventeenth century, when Kepler demonstrated that planets move in elliptical orbits around the sun, that the Copernican system became noticeably simpler than the Ptolemaic.

While it was Kepler's work on the elliptical orbits which ultimately converted working astronomers to the Copernican view, it was his contemporary Galileo who made the public aware of the debate. His discoveries with the telescope were astonishing, and his writings were entertaining and polemical; Copernicus had written for mathematicians, but Galileo's books were intended for a much wider readership. His view of mathematics was quite different from the traditional one; for he believed that mathematics was the language of nature, and that mathematical reasoning could bring certainty into physics. This idea was not altogether new, for it was associated with the Platonic tradition; and in antiquity, Euclid had handled problems in optics as matters of applied geometry. The greatest example was the work of Archimedes in the third century BC. We know about his leaping from the bath and crying 'Eureka', but his own treatise on specific gravities mentions neither this, nor crowns, nor even any experiments. He sets out an austere definition of a liquid, and calculates what will happen when objects of various shapes and densities are immersed in such a medium. Perhaps more striking is his work on the lever, because he demonstrates how point masses can be balanced using rigid weightless rods; in contrast to a rather earlier Aristotelian author who warns his reader against the tricks that traders may play upon him with actual unfair balances. Archimedes thus considers an ideal world; real liquids and real balances or levers approach more or less closely to his ideal, mathematical constructions. But his reasoning seems to have a certainty that a generalisation from a number of actual baths or balances could never have; just as a proof of Pythagoras' theorem has an authority far beyond the result of measuring a number of right-angled triangles.

Archimedes' work became well-known again in the sixteenth century; and Galileo could be described as his disciple. In antiquity and in the seventeenth century, geometry was particularly esteemed because it seemed as though the geometer could solve practical problems by pure thought. To prove Pythagoras' theorem does not require that one is familiar with countless triangles, or indeed with any. The axioms of Euclid's system say nothing about the world, apparently, and yet from them one can deduce this and all the other theorems, which can be used to enable us to lay out or survey towns or fields. It seemed as though the geometer had the key to the understanding of nature; or in Galileo's phrase, that the book of nature was written in the language of mathematics.

By the nineteenth century, this belief required modification because of the invention of non-Euclidean geometries. One of Euclid's postulates can be stated as 'given a line and a point, only one line in the same plane can be drawn through the point parallel to the first line'. This seems sensible enough; but if one supposes instead that several lines, or none, can be drawn through the point parallel to the first one, then a consistent system of geometry can be drawn up in which the theorems differ from those in Euclidean geometry. And with the coming of Einstein's theory of relativity, it became an open question whether Euclidean geometry or a non-Euclidean geometry was best suited to physics; and indeed he opted for a non-Euclidean one. This is a very general case of the problem of mathematical physics; there may be no doubt that an arrangement of weight-less rods would behave as Archimedes says it would, but such an arrangement may be so far from anything in the real world as to be valueless. To us it is usually evident that the applica-bility of the mathematical model must be tested; in the early seventeenth century this would have been apparent to Aristotelians, but less so to those in the Archimedean tradition to whom the power of mathematics was obvious.

There were therefore two possible routes which seemed to lead towards certainty in the seventeenth century. One involved the search for mechanical explanations, nature being assumed to be a vast mechanism like an overgrown clock; while the other entailed the search for simple mathematical expressions which must lie behind the phenomena, God's rationality guaranteeing the rationality of the world. The great triumph of the mathematical approach to nature was Galileo's discovery of the law of falling bodies; he took it for granted that this accelerated motion must be the simplest kind of accelerated motion, and did some rather perfunctory experiments to show that this was so. This law implies that all bodies would fall at the same rate, which is manifestly untrue; but Galileo invoked air resistance to explain why feathers fell more slowly than lead. His belief in simplicity seems to have lain behind his failure to appreciate the discovery that planets move in elliptical orbits, even though Kepler had sent him a copy of his book; for to

Galileo circles were a simpler curve, and the divergences from circles must be due to some force which produces a distorting effect just as air resistance masks the simple law of falling.

In Newton, whose *Principia* appeared in 1687, the two traditions found their culmination; though the mathematics alone seemed to bring certainty, while the search for mechanical explanations seemed to lead into the realms of hypothesis and query. Newton was able to establish that Descartes' mechanism of vortices would not maintain planets in the orbits in which they are observed to travel; but that these orbits were to be expected for bodies which attracted each other with a force proportional to the inverse square of the distance between them. He and Halley were able to predict the motions of comets, which had hitherto seemed random; and to account for the variations in the orbits of some of the planets. Because Jupiter and Saturn, for example, attract each other, when they happen to be relatively near, with a force which cannot be neglected in comparison with the sun's attraction on each of them; they pull each other slightly out of the elliptical orbits that a single isolated planet would follow. Newton, starting from Kepler's laws, was thus able using his much more general theory, to account for divergences from these very laws.

His greatest achievement was to link terrestrial and celestial physics, by showing that the moon was falling just as apples fell. But the problem seen by contemporaries in looking at his work was that it seemed to require action at a distance. That is, it was mathematically elegant but physically implausible. This is a charge that has frequently later been brought against applied mathematics in the sciences, and is closely parallel to the charge brought against it in history and sociology today. In Newton's case the problem was that pushes and pulls are comprehensible, but that it is hard to imagine how two bodies could attract each other across empty space. Newton had shown that the vortices of subtle matter envisaged by Descartes would not work; and his own toyings with an æther came to nothing, because the æther would have to be thick to propel bodies towards each other and thin so as not to interfere with inertial motion. To his critics, therefore, it seemed as though Newton's theory of gravity involved the perpetual miracle of attraction across void space.

46

It also seemed like a backward step. Mechanical philosophers had just escaped from a kind of science popular in the Renaissance which was called natural magic. For the natural magician, the world was full of sympathies and antipathies, and action at a distance was to be expected. Man, the little world or microcosm, was an analogue of the great world or macrocosm. Events in the heavens would therefore be paralleled by events in men's lives; and astrology was not unreasonable.

The sixteenth-century mystic and doctor Paracelsus believed that the processes which in alchemy would convert base metal into incorruptible gold would cure diseases in man. He therefore introduced mercury into the pharmacopoeia; and found indeed that it was a specific for syphilis. Later attempts with other metallic drugs were less happy; but a school of physicians followed the general doctrines of Paracelsus. In the early seventeenth century some of these proposed a new cure for wounds, in which the weapon was smeared with an ointment or salve while the wounds were simply bound up with clean bandages. It was found that patients thus treated did usually get better sooner than those whose wounds were treated with ointments. To the Paracelsians, it seemed as though weapon-salve at a distance was more effective than ointment directly applied; orthodox doctors could not believe that, and denied the cures. We can laugh at both sides in the controversy; but it was episodes such as this which made any explanation involving action at a distance seem thoroughly dubious to a sceptical natural philosopher.

The science of the natural magicians was closely tied to practice; to making predictions, curing the sick, or converting metals into gold. But in this context, what is more interesting is that it was irrefutable. If an astrologer, for example, made a prediction it might be so couched as to be unfalsifiable, as the pronouncements of oracles have always been. But if it was falsifiable, it might turn out to be true; then the astrologer would of course gain prestige. But on occasion it would turn out false. This would not discredit astrology, any more than a false prognosis discredits modern medicine. The astrologer might admit that he had made a mistake; or he might say that obviously some planet whose influence he had thought it safe to neglect because it was small had in the event tipped the balance;

and he might add that this failure showed the need for much more research in astrology. Kings and Emperors in Europe in the first half of the seventeenth century employed eminent astronomers as their astrologers, and the astronomers cast horoscopes in which they seem to have at least half believed. During the Civil War and Interregnum in England an enormous number of treatises on natural magic were published; both sides consulted astrologers, and on occasion the same one.

It was not because of inaccurate predictions that eminent and educated people stopped using astrologers as counsellors, but because of the success of the natural philosophers like Descartes and Boyle who insisted on a mechanical world picture in which the actions at a distance involved in magic were unacceptable. The explanations given by the astrologer or by the physician using weapon-salve were no longer credible, even if their predictions or cures were successful. Men of science have never long been satisfied with pragmatism.

Newton was not satisfied with his theory, for he believed that action at a distance was impossible. But his disciples came to accept it; and Roger Cotes, who edited the second edition of the *Principia*, wrote sarcastically about those who demanded a mechanical explanation in addition to the equations which Newton had provided. For him science was thus a matter of exact laws linking phenomena, and to explain a phenomenon was to state these laws rather than to invoke a hypothetical mechanism. The mathematicians' science involved quantitative relationships between phenomena, while the magicians' had been qualitative; but neither insisted on a physical account of the connections between events. This was the version of Newtonian method which during the eighteenth century became characteristic of the French; and it is associated with Newton's famous remark '*Hypotheses non fingo*', I do not feign hypotheses.

But this remark had its context, in the 'General Scholium' appended to later editions of the *Principia*, and should not be taken as the essence of Newton's method. For most notably in his *Opticks*, which appeared in 1704, he showed himself a master of the experimental and inductive way of proceeding; and in the 'Queries' which he added to later editions of this book he set out his guesses about the mechanisms that lay behind phenomena. It is here that we find his famous speculations about atoms and particles of light; and natural philosophers in England avidly read these 'Queries' rather than the arid mathematics of the *Principia*. Indeed, it seemed as though Newton had satisfactorily accounted for the facts of astronomy, leaving in the *Principia* an almost-completed edifice; while the 'Queries' seemed to be his legacy to the next generation, being problems not yet solved and mechanisms not yet established. Even a hundred years after the death of Newton in 1727 these 'Queries' were still turned to as an authority in Britain; and today they have a fascination which is absent from the formal 'Rules of Reasoning' which Newton set out in the *Principia*.

The mechanical and mathematical world views carried all before them in the early eighteenth century, and it became accepted that a properly scientific explanation should involve an acceptable mechanism and be quantitative. This pattern did not happily fit either chemistry or biology; and indeed it seems that we can say with benefit of hindsight that both these sciences made slower progress because their practitioners were too dazzled by the successes of the physicists. In chemistry it proved impossible to apply atomic ideas to give a detailed and falsifiable account of any phenomenon before the latter part of the nineteenth century; and interatomic forces proved much more intractable than celestial ones. In Newton's mechanics, every particle of matter attracts every other particle; but in chemistry we meet with elective attraction—some substances readily react together, and others do not. Again, it was not until the mid-nineteenth century that a coherent account of chemical dynamics could be given. There were critics of the 'Newtonian' paradigm in chemistry in the eighteenth century; thus G. E. Stahl wrote that it did not get to the kernel of things, or in our terms that it only provided explanations in principle. His theory of *phlogiston* can be seen as a groping towards a concept of chemical energy, but phlogiston was taken to be a queer kind of material substance, and as such was exploded by Lavoisier in the 1780s. Lavoisier himself played a role in the separation of chemistry from physics, for he urged his readers to avoid any speculations about atoms, which could only divert them into metaphysics.

Chemistry at the end of the eighteenth century thus became

on the one hand a fundamental science, for physiology, electricity, and mineralogy seemed to depend upon it; and on the other a science without a general mechanical background or a satisfactorily quantifiable basis. There were some laws of chemical combination, but Dalton's atomic theory seemed to most chemists of the early nineteenth century crude and lacking in a dynamical and fully quantitative aspect. Chemistry therefore in what can be called its golden age, when it was full of vitality and attracted a great deal of attention, did not measure up to the 'Newtonian' criteria; and chemists sighed for another Newton who would reorganise the science into a proper pattern, as Copernicus, Kepler and Newton had organised the mass of astronomical knowledge.

In biology the methods of physics worked no better. Although Descartes was much less of a biologist than Harvey, it was his mechanical ideas, pointing towards *'l'homme machine'*, that governed much of the physiology of the early eighteenth century. Cartesians and Newtonians measured blood pressures as they measured air pressure; and sought to account for the body's heat as produced by some kind of internal friction. With the rise of chemistry came a successful reduction, when the function of respiration was explained in the 1780s by Lavoisier and Crawford as a kind of combustion. They believed that oxygen was absorbed to burn food and maintain the body temperature; and thus explained why we need to breathe, which had previously been a mystery—the usual view had been that breathing was a means of keeping the heart cool. Despite this success, biologists began to be unhappy about following Newtonian paradigms; and John Hunter in Britain and then Cuvier in France stressed the unity of the organism, and the weakness of the analogies between animals and machines. Hunter was particularly struck by the observation that after death the stomach is sometimes digested by the juices it contains. The digestive juices therefore attack dead but not living matter; physics and chemistry are sciences of dead matter, and can elucidate the cause of death, but cannot form the basis of the life sciences.

The work of Hunter and then of Cuvier marked a return to teleology; and, since the eighteenth century was also the great period of classification in natural history, the biological sciences generally saw a return to the ideas of Aristotle. John Ray in

Britain had attempted a natural classification at the end of the seventeenth century; then Linnaeus had produced a convenient artificial one, based upon counting the sexual parts of flowers, for botany; and by 1800 botanists in France were developing a natural system, based on multiple criteria, to avoid the anomalies resulting from Linnaeus' method.

By the late eighteenth century, it was clear that the mechanical and mathematical paradigm would not fit some of the most important sciences; that there were a range of different kinds of explanations, which would be appropriate in different situations. Those dedicated to the mechanical programme could only declare, as they still do, that sciences where it did not work were 'at the natural history stage' as though this were an awkward age through which all sciences had to pass; but this kind of historical determinism need not be admitted.

Cuvier used his teleological framework of ideas in sorting out the jumbled masses of fossil bones which were turned up in the Paris Basin as it was quarried to rebuild Paris as an imperial capital. He and his contemporaries on both sides of the Channel revealed in the first two decades of the nineteenth century the succession of living forms that had peopled the globe, and most of which had now become extinct. The static view of biology characteristic of the eighteenth century had to give way to a dynamical view; first of the history of organisms, in physiology, and then (in the work of Charles Darwin) of the relations between organisms. With palaeontology, the study of fossil organisms, historical explanation entered the sciences. The timeless 'best of all possible worlds' of the eighteenth century had to be given up in favour of a world which had come to its present state through a whole series of other states.

At first it was generally assumed that the transitions between these states had been sudden, and the past history of the earth was seen as a series of peaceful epochs separated by cataclysms; rather as a naïve view of human history discloses peaceful periods separated by revolutions. But in 1830 Charles Lyell began to publish his *Principles of Geology*, in which he argued that past changes should be accounted for in terms of forces in action at the present. His predecessors had been prodigal of violence and parsimonious of time; the catastrophes which they had posited were unnecessary if nature were allowed enough millions of years to bring about the changes disclosed in the rocks. His view gradually prevailed; and his greatest disciple was the young Darwin, who in 1859 applied the same historical method to understanding the variety of creatures which we see around us. Darwin thus explained the natural system of classification; animals or plants which show a great number of similarities in structure, or homologies, are descended from a common ancestor in the not too distant past. In this historical kind of science, to explain something is to describe its antecedents. In the years after the *Origin of Species* was published, this kind of explanation became very widely popular throughout the sciences; and attempts were made to trace the development of chemical elements, and the evolutionary history of stars and of planetary systems.

This was not the only new kind of explanation to enter the sciences at this time. For in the same autumn in which the *Origin of Species* was published, in fact on 21 September 1859, the great physicist James Clerk Maxwell read his first paper on the kinetic or dynamical theory of gases to the British Association. It was not an altogether new idea that a gas might be composed of particles all rapidly moving hither and thither, and colliding with each other and with the walls of the container and thus exerting a pressure. What was new in Maxwell's treatment was that he handled the problem statistically; and this was the first statistical explanation in physics.

Statistics was not a new science. It goes back to the early eighteenth century, to the early days of insurance on the one hand, and the study of games of chance on the other. Halley computed life-tables based upon data from Breslau (where the records were better than from English cities) in order to assist those offering annuities; for in the business of life insurance one must know the expectation of life at different ages. Clearly, different people will die at different ages, but the object in life insurance is to know the average age at death; and this is determined from data assembled in respect of numerous people, none of whom may have happened to die at the average age. Here the computations are improved as more information is collected; the basis of this kind of study is clearly empirical.

Games of chance can be handled in a more deductive manner. One cannot predict the life span of a man without knowing a good deal about him and about men in general; but one can with much less information predict that a coin should come down heads as often as tails when it is tossed—though without experiments one could not be absolutely certain that it would not go upwards for ever, or land and remain on its edge. The behaviour of dice is similarly predictable; and in the study of games of chance of this kind, one handles the coins or dice as ideal discs or cubes just as Archimedes had discussed levers in terms of weightless rigid rods. But whereas his conclusions had been definite, that a certain weight at a certain distance from the fulcrum would balance another weight at another distance; the conclusions of the analysis of a game of chance are statistical, that a certain combination of throws can only be expected to happen once in so many times.

The great astronomer Laplace, who briefly held high office under Napoleon, tried to apply this kind of deductive statistics to society. In considering the ideal number of men to form a jury, and the majority needed to secure right verdicts, he began not with empirical information but with assumptions about the frequency with which men tell the truth and make sensible judgements, and about what are acceptable rates for the conviction of innocent people and the acquittal of guilty ones. It seems that this was intended in all solemnity as a contribution to social science rather than an amusing exercise; and to his critics even within the sciences Laplace came to stand for the deductive reasoner who lacks judgement outside the mathematical field.

It was Gauss, his younger contemporary, who brought statistics out of life insurance and games into the mainstream of science; but as a tool rather than as a basis for explanation. To him we owe the famous normal distribution curve, on which the heights, weights or examination marks of a fair sample of people should for example fall. He developed it not with this in mind, but in the course of an attempt to estimate the errors in astronomical observations; and the curve was indeed at first called the error curve. If one repeats any observation, one gets a slightly different reading as anybody who has taken bearings, or done titrations in chemistry, will know. Astronomers down to the end of the eighteenth century averaged their observations, usually leaving out any which seemed particularly discordant and therefore bad. This seemed reasonably satisfactory when it involved one observer, or even one observatory.

But in the second half of the century there came a great surge of observation all around the world with the transits of Venus across the sun in 1761 and 1769. About once a century there comes a pair of such transits, when the planet passes across the face of the sun; and Halley had shown that observation of the transit from places far apart on the earth could provide data for computing the distance of the earth from the sun. Newtonian dynamics and Kepler's Laws only enabled relative planetary distances to be determined; and attempts to measure the sun's distance by any direct method could not give an accurate figure. Because of the importance of astronomy for navigation, governments supported expeditions to observe the transits; and observers were also ready at the great observatories.

The problems were that the exact moments at which Venus began to pass across the sun's disc, and left it, turned out to be hard to observe; and that there was too great a wealth of information for astronomers to handle. The various observers differed in skill and in the favourable or unfavourable position of their instruments, which also differed in excellence. Astronomers at the observatories of Europe had to try to weight the various contributions made by different observers; and probably paid most attention to observations sent in by those they knew something about. But the final result obtained depended upon which observations were given most weight; and therefore there was no clear-cut result for the earth-sun distance at the end of it all. The error curve helped astronomers reducing observations to arrive at the true value, taking discordant observations into account; though even Gauss could not arrive at an unambiguous result from the eighteenth-century transit observations.

Quetelet in Belgium in the mid-nineteenth century extended the statistical methods he had learned in astronomy to the social sciences. He showed how predictable human behaviour was in the mass, and thereby gave comfort to determinists. People commit murder for all sorts of reasons; and yet Quetelet showed that the murder rate was surprisingly constant. He could not predict who would murder or be murdered in the coming year; and yet he could predict how many murders there would be.

It was this statistical work which Maxwell adapted to fit his model of a gas. Since the particles are all colliding with each other and the walls of the container, they will all be moving at very different speeds; at any instant some indeed will be stationary. Since Maxwell knew that he was dealing with very big numbers of particles, he knew that their velocities would be distributed about a mean according to Gauss' error curve; and he found on making this assumption that the properties of a congeries of particles fitted well those observed for a gas. He found that in his model the viscosity or stickiness would be independent of the pressure which the particles were exerting; and verified this surprising conclusion for actual gases by experiment.

Maxwell showed that some paradoxes would follow if one supposed a being—his famous demon—small enough to discriminate actual particles. Such a creature could by working a trap door between two vessels collect the fast-moving particles on one side and the slow-moving ones on the other. He would thus, without expending any energy, have produced a hot and a cold vessel of gas from what had originally been two tepid ones; and this result would defy the second law of thermodynamics. The kinetic or dynamical theory of gases thus depended upon the number of particles being large, and it might lead to paradoxes if one could consider individual particles.

Maxwell was the founder of the Cavendish Laboratory at Cambridge; and his most eminent successor there, Rutherford, found himself forced to extend statistical explanations. Radioactive substances were found to decay at definite rates, so that half the radium in a given lump would always have decayed in a set period. But there seemed to be no way in which the particular atom which was on the point of decaying differed from its neighbours; and no chemical or physical process which would speed up or slow down the decay. Thus there seemed to be a process to which no definite cause could be assigned, namely the decay of a particular atom; but which, given a large number of atoms, was extremely regular. There seemed to be law but not cause in the microsphere. As one can predict how many murders there will be without being able to predict who will commit murder, so it seemed that the laws of radioactive decay rested upon statistical grounds. Men of science were not all happy with this situation, but the search for hidden parameters was fruitless and for a time at least statistical explanations have had to be lived with.

In place of any one notion of scientific explanation, we have thus found a whole series of them; explanations in science may be teleological, mechanical, mathematical, historical, or statistical, and no doubt there are other kinds also. It is rare for everybody working in a science to be satisfied with the kind of explanation then in vogue in his discipline, except in the dullest and most 'normal' periods of history. But when attempts have been made to apply in one science notions of explanation which have proved appropriate in another, the result has been frequently unsatisfactory. Mechanical explanations in chemistry and physiology in the eighteenth century failed, as we saw, to open a path to fresh discoveries, to account for any phenomena in detail, or to help those learning the subject; and these three objectives should be in the mind of everybody proposing an explanation. It does not follow, in the natural and no doubt also in the social sciences, that to model one's explanations upon those currently fashionable in physics will lead to progress.

Within a science, a certain kind of explanation may be popular in one country, and another kind elsewhere. Thus about 1725 most astronomers in Britain were satisfied with Newton's mathematical account of the solar system; while in France and Germany they demanded an intelligible mechanical account. A hundred years later, the boot was on the other foot; Laplace and his contemporaries had proved the power of mathematical analysis in solving a whole range of problems, while in Britain mathematics had been neglected and the newest techniques were little known. It was only in the 1820s that mathematics at Cambridge was brought up to date through the efforts of a few energetic young men.

When electromagnetism was discovered by Oersted in 1820, the greatest investigators of it to emerge were Ampère in France and Faraday in England. Ampère was an able mathematician, and he could account for the discoveries made by Faraday and others; Faraday was ignorant of mathematics, but had a strong feeling for physical plausibility. While Ampère worked with equations, Faraday had a mental picture of lines of force filling space; and this picture enabled him to visualise new situations, to make discoveries, and to reject Ampère's theory as a real account of the phenomena. Maxwell was in the end able to put Faraday's insights into mathematical form; but he, and the Cambridge mathematicians active in the nineteenth century generally, retained a strong feeling for physical reality. In Victorian Britain, it was generally accepted that every variable (x or y) appearing in an equation in a piece of applied mathematics should be assigned some empirical significance. On the Continent, the feeling was much stronger that science was a matter of laws linking observables, and that the important thing was to make accurate predictions.

Similar differences in taste can be found between scientists in different fields from physics; and even between graduates of different universities in the same country. Perhaps today, with more mobility between universities and more international conferences, these differences may be diminishing; but international communication is never perfect. Again, as science has become more specialised, the differences in world view and thus in the kind of explanation found satisfactory between scientists in different disciplines have increased.

Through the history of modern science we have therefore had not a single pattern of scientific explanation which can be contrasted with a similar single pattern to be found in non-scientific disciplines, or in ordinary life. We find instead a family of explanations, having something in common but not all reducible to one kind. This means that in the relations of the sciences to other fields of knowledge, we do not find a monolithic science confronting the arts or social sciences. We find a complex relationship, with borrowings both ways and occasional conflicts. It is to these relationships, to science as an intellectual activity in its relations with other intellectual activities, that we shall turn in the next chapter.

3
SCIENCE AND NOT SCIENCE

In some languages, the word equivalent to the English word 'science' covers any organised body of knowledge; and therefore history and philosophy are sciences just as chemistry and physics are. In English, that is no longer the case; and therefore in discussing the relationship of science to what is not science, we shall not be contrasting the rational with the irrational, or genuine knowledge with mere guesswork. To say that history is not a science is not to say that it falls short of some ideal, exemplified perhaps by physics; but to say that its objectives and methods differ sufficiently from those of the physical or biological sciences to make it sensible for us to put it into a different category. We shall be using 'science' then in its usual sense, to mean the physical, biological and geological sciences, and to include engineering and mathematics; that is, the 'natural knowledge' which the Royal Society was founded to improve.

In this chapter, we shall be investigating some of the relations between science and other intellectual activities; some of them less and others more rigorous than the science of the day. Since there has never been a single unified science, there have always been complex relations between different sciences, and between sciences and other activities. And because it makes little sense to talk of scientists as though they formed a separate class or profession before the nineteenth century was well-advanced, we must remember that men of science or natural philosophers were also, or primarily, physicians, country gentlemen, clergymen, industrialists, and so on. The word 'scientist' was coined in the 1830s and only slowly came into use as the number of those whose chief concern was science grew. Sometimes those whom we remember as men of science were famous in their own day for some quite different reason; or conversely were harassed, not as scientists but as religious dissenters, tax collectors, or debtors. Not every conflict in which men of science have been involved had to do with science.

One of the more complex interactions has been that between the sciences and theology. In the late nineteenth century, this was sometimes described as a kind of warfare, in which science had always proved victorious even over the combined forces of Church and State. This imagery tells us something about the views on religion of late Victorian publicists of science; but as a guide to what went on in the nineteenth century or earlier it is very poor. It was only in the nineteenth century, when the Church had no power to persecute any but its own members, that it became at all plausible to argue that theology was dealing with a 'God of the Gaps'; that is, that it was only where other kinds of explanation had so far failed that God was invoked. The relationship was really more interesting than that.

In Plato's *Laws* and in Boethius' *Consolations of Philosophy* can be found examples of natural theology; that is, attempts to demonstrate the existence and benevolence of God from nature. The standard arguments were based, in antiquity and in modern times, on the regular movements to be observed in the heavens, and on the coherence of the parts of living things. While the traditional Christian teaching had been that the Scriptures, interpreted by the Church, contained all that was necessary for salvation; the new philosophers of the seventeenth century urged that our knowledge of God began with the study of nature, and that His revelation in Scripture was a supplement to this natural religion. The study of nature was therefore a religious duty; and authors like Francis Bacon wrote of the two books from which we could learn about God, the Book of Nature and the Bible, which complemented each other.

In Bacon's time the first major crisis in the modern period was developing over the Copernican system. Copernicus' book had appeared in 1543 in response to pressures from both astronomers and ecclesiastics, and with a dedication to the Pope; it also contained an unsigned preface added by the publisher which stated that the author's objective was only to save the appearances, which was not true. The Council of Trent, which met between 1543 and 1563, inaugurated the Counter Reformation; and brought about an assertion of Church authority and a suspicion of new ideas which had not been evident during the preceding century. In the later sixteenth century, the Copernican system found favour with a number of astronomers; but the most notable convert was Giordano Bruno. Bruno was a Dominican who had left his order, and was imbued with pantheistic, Hermetical, and magical ideas; indeed it seems that he hoped to replace Christianity with a form of sun worship. He approved of Copernicus' work because the sun appeared in its rightful place as the centre of the system; but was scornful of narrow-minded astronomers who had failed to see the astrological and magical value of the Copernican diagrams. After denouncing pedants up and down Europe, Bruno fell into the hands of the Inquisition in 1593 in Venice, and was burnt in 1600. Given an Inquisition with power to punish heretics, there can be little doubt that in Bruno they had found an appropriate victim; and little doubt too that he cannot really be seen as a martyr of modern science.

But Bruno did show that scientific ideas taken too seriously could lead to dangerous thoughts; and Cardinal Bellarmine, who had dealt with Bruno, soon afterwards found himself faced with Galileo. In 1610 Galileo published his little book the *Starry Messenger,* or *Message,* in which he described his discoveries with the telescope. He had seen the mountains on the moon, which therefore was a body not unlike the earth although the received view was that it was a perfect sphere of quintessence; numerous new stars, showing that the universe was much bigger than had been supposed; and the moons of Jupiter, which clearly did not orbit the earth. In Galileo's opinion, these facts—and the phases of Venus, which he observed soon afterwards and which showed that Venus goes round the sun rather than the earth—established the Copernican system as a physical reality.

His observations were soon confirmed by the Jesuit astronomers at the Roman College; but his interpretations were less acceptable. Conservative astronomers by this time no longer believed that all planetary orbits were centred on the earth anyway; they adhered to the view of Tycho Brahe, a generation younger than Copernicus, who held that the earth was stationary and that the sun went round it, all the planetary orbits being centred on the sun. This is geometrically equivalent to the

Copernican system, and no simple observation could prove one rather than the other; it also seemed closer to common sense to have the earth at rest.

Again, on the Copernican system the stars had to be an enormous distance away. If the distance of the fixed stars were comparable to distances on earth, then one would expect that as one moved about the stars would seem to change their positions; such a change would be called stellar parallax. No such parallax is seen, and the fixed stars were therefore known to the Ancients to be so far away that the earth was a mere point compared to the heavens. But on the Copernican system, and earth is supposed to be moving in a vast orbit about the sun. One would predict therefore that there would be parallax if one observed in January and then in July, when the earth would be many millions of miles from where the first observation was made. No parallax could be observed (until the nineteenth century, when telescopes were very accurate), and this implied that the diameter of the earth's orbit was insignificant in comparison with the distance of the stars; or else that Copernicus was wrong.

To Bellarmine, it seemed that at most Galileo could argue that the Copernican system was convenient; but that was not enough for Galileo. He wanted to get away from a physics of likely stories. He was informed in 1616 that the Church required that the Copernican system be neither held nor taught; but found this restriction galling. On the election of a new Pope, he got permission to write a book comparing the two systems and duly published it in 1632. It was hardly a fair comparison; the obsolete Ptolemaic rather than the Tychonic system was compared to the Copernican, and Galileo could not strictly prove that the Copernican system was true, though he could make fun of adherents to the Ptolemaic one. His wit had already made him many enemies; and this time it got him into the hands of the Inquisition, which meant that he had to face trial and then spend the rest of his life under house arrest.

Despite Galileo's condemnation, the Copernican system was taught—as convenient rather than true—even in Roman Catholic institutions in the seventeenth century; and the work of Kepler and then of Newton made it increasingly hard to doubt that the earth went round the sun, despite the lack of stellar parallax. Galileo's conviction that science was a matter of truth rather than probability rapidly gained ground, as people sought certainty in a blend of mathematics and mechanics. The Inquisition had won a battle, but lost the war; for the sciences had achieved by the middle of the seventeenth century an autonomy which they had previously not had. This did not mean that in the future new scientific ideas were always welcome, but it did mean that conservatives (who were naturally enough often right in resisting innovations) could no longer call upon formal coercive machinery to put their opponents to silence.

It is sometimes implied that from the time of Galileo's condemnation onwards, science forsook the Roman Catholic part of Europe and went north. Certainly economic dominance in Europe passed from the Mediterranean countries to those bordering on the Atlantic and the North Sea, but Galileo can have had very little to do with that. Since France was a leader in science as in other cultural activities, Roman Catholicism cannot have had a necessary connection with scientific stagnation; and while Italy did not perhaps produce another scientist of the stature of Galileo during the seventeenth century, there were there a large number of very able men of science, including Torricelli, Malpighi, and Borelli.

What does seem to be true is that natural theology began to flourish in the northern part of Europe, and particularly in England. Theologians increasingly came to welcome the new science as an ally rather than to see it as a threat. Galileo had perhaps unwisely quoted a witty cardinal as saying that the Bible was to tell us how to go to Heaven and not how the heavens go; and he had used the authority of St Augustine to support his view that where a statement in the Bible appears to conflict with empirical knowledge, then it is to be interpreted in an allegorical sense. This meant that empirical science was an arbiter in biblical interpretation. Because of his remarks on predestination, Augustine was a popular authority among Puritan theologians; and Galileo's arguments were taken up with enthusiasm by John Wilkins in works published from 1638 on. He became Warden of Wadham College, Oxford, during the Interregnum, and married Oliver Cromwell's sister;

after the Restoration he became Secretary of the Royal Society, and then Bishop of Chester.

Wilkins was chiefly interested, following Bacon, in demarcating the spheres of theology and natural philosophy; but after the Restoration of 1660 came the great flowering of natural theology. One might have expected that theologians would have been concerned with accounting for the pain and imperfection which seems evident in the world; and indeed Thomas Burnet in his *Sacred Theory of the Earth* did seek to show how the earth was a ruin of that upon which Adam and Eve had been created. At the great cataclysm of Noah's Flood, the previously smooth crust of the earth had been deformed into mountains and seas, while its axis had been tipped bringing to men the vicissitudes of seasons. Burnet tried to give naturalistic explanations of this, and of the restoration of the original state of affairs by fire which is promised in the Bible.

Most of Burnet's contemporaries sought on the contrary to demonstrate that this was the best of all possible worlds; but they agreed with him in putting evidence from Genesis on a par with that from scarped cliff and quarried stone. Biologists like John Ray and William Derham produced the traditional teleological arguments for a designer of the universe; but they emphasised also the variety to be found in plants and animals. This was even more evident by the seventeenth century than it had been in antiquity; for the fauna and flora of America, Africa, and Asia had turned out to be very different from that of Europe even where the climate was similar. It seemed that while mechanical processes would have produced identical creatures in similar situations; only a benevolent creator would have given us such diversity.

The new astronomy of Newton also provided ammunition for the natural theologian; for in place of the awkward machinery of spheres and epicycles, astronomers had revealed the one simple law of gravitation which lay behind all the celestial motions. Other branches of science which seemed chaotic would no doubt be found to be equally simple when looked at aright. Even the vast extension in the size of the universe which the Copernican system entailed (because of the absence of stellar parallax) was to be welcomed, because an infinite God would be expected to create an infinite universe rather than one which was just a few million miles across. There were also some functions reserved for God in the Newtonian world. He had to keep up the quantity of motion in the system, so that the celestial clock would not run down; and to ensure that the attraction of the planets for one another did not eventually pull them out of their orbits. This aspect of the Newtonian system was distasteful to Leibniz, who felt that it involved saying that God had made a rather poor piece of clockwork which needed to be wound up and set from time to time like a human artefact. Newtonians retorted that a God who never interfered in the universe would be like an absent prince, whose subjects would soon find that they could get along very well without him; but Leibniz's conception of a creator who did not need to intervene in the world, because He had foreseen everything that could happen and allowed for it, was one which prevailed very generally in the eighteenth century, and was called Deism. Natural theologians tended to be bland in their discussion of evil in the world—after all, if this is the best of all possible worlds then there is nothing much that can be done about anything; and rosy in their view of the animal world, in which they saw carnivores as a device for bringing euthanasia to elderly herbivores. They also tended to be condescending towards God, congratulating Him where it seemed appropriate on a neat bit of work in the design of some reptile or insect.

These aspects of natural theology affronted some theologians by about 1800; notably Schleiermacher in Germany, who believed that religion should be based upon feeling and not on alleged proofs; and Coleridge in England, who urged clergymen to make men feel their want of God rather than to try to prove His existence. For the 'First Cause' or 'Creator' of the natural theologians was far from the personal God of Christianity; an intellectual acceptance of His existence would have had little effect upon anybody's life. Moreover, the arguments for the existence of God from nature or logic had come to seem distinctly shaky; Hume in Scotland and Kant in Germany had indeed shown the unsoundness of all of them. Hume's work on natural religion is particularly interesting because in some respects he anticipates the Darwinian view of nature red in tooth and claw; but he was not interested in biological theory so much as in showing that nature was neutral. We might see benevolence and design, but we could just as easily see nature as demonic, or blind, pouring forth from her womb her maimed and abortive children into a world in which the strong terrify the weak and disease is everywhere. Nature only discloses God to those who are already looking for Him, and know how to recognise Him.

But if natural theology was unsound, both as theology and logic, it had the value of connecting science with a system of beliefs and an ethical code. Science and technology are morally neutral, and too much should not be expected of them; but in some places science did become a kind of ideology, the assumption being that if the sciences were generally known and widely applied then people would become happy and good. The disappointment of this chimerical hope must be one of the factors which lies behind the so-called anti-science movement. Where, as in Britain in the eighteenth and much of the nineteenth century, science was usually taught as natural theology, it was related and subordinated to other human concerns, and could not become an ideology. We cannot return to this kind of natural theology, but we can be sceptical about wild claims for science, and not allow it to claim some kind of mystique.

It should not be supposed that all natural theologians agreed with one another. In the 1770s Joseph Priestley, a Unitarian minister and the discoverer of numerous gases, seized upon the new atomic theory of the Jesuit Roger Boscovich, according to which atoms were mere points surrounded by fields of force. This to some extent overcame the problem of action at a distance in the microsphere, for atoms did not have to push and pull each other—their spheres of attraction and repulsion simply interacted, and space was everywhere full of forces. The older billiard-ball atomism had been used by seventeenth-century theologians, such as Ralph Cudworth, to support belief in immaterial substance; for atoms of passive, inanimate brute matter could not even stick together without something else to hold them, and still less could they form living and thinking beings. Man cannot therefore be composed of mere matter, but must have an immaterial soul; and if the atoms of matter cannot be destroyed, then still less could the soul be.

Priestley believed in the resurrection of the body rather than the immortality of the soul; and thought that because the new atomism of Boscovich involved active atoms—for each was associated with spheres of force—it did not need immaterial substance. If souls were an unnecessary entity, then they should not be postulated; and Priestley believed that his materialism was closer to primitive Christianity than was the belief in the Trinity and in immortal souls, both of which seemed to him to be corruptions introduced from Plato's philosophy. Just as the atomists of a hundred years earlier had been at pains to dissociate themselves from the alleged atheism of Hobbes; so those of late eighteenth-century Britain were careful to disclaim belief in materialism, which recent events in France had shown to lead to civil disorder. Priestley had duly supported the French Revolution as well as dynamical materialism. It was not until well into the nineteenth century that atomic theories in physics or chemistry could be discussed without this spectre of materialism being raised. As the concepts and language of the physical sciences became more technical and less accessible to the layman, so their general implications became less evident and less pressing.

By the early eighteenth century it seemed as though there was no serious conflict between the theological and the scientific views on ,astronomy. Rather they seemed complementary; astronomers produced evidence of design and wisdom in the heavens, while theologians tried to relate these to the great problems of suffering, death, and redemption. It no longer caused any general unease that the account of the creation of the sun, moon and stars given in Genesis did not square with that given by the astronomers; though this was occasionally pointed out by anti-clerical authors. A number of astronomers were clergymen, particularly in the universities where the majority of those teaching were in holy orders; but they did not expect to find a theory of astronomy set out in biblical texts, and accepted that even an inspired author would adapt any references to astronomy to the understanding of his contemporaries.

The eighteenth century was also a great period for natural history; but because in Genesis there is rather more about the creation of the earth than of the heavens, its authority as a source of information in geology, botany, and zoology survived.

It survived in other fields too; anthropology for example began under the shadow of the Tower of Babel, and the various races of men and their languages were traced back to events described in Genesis. To do this nowadays would be fanatical; but in the eighteenth century it was normal, and men saw no clash between accepting evidence from the Bible and from the rocks. It was not that they refused to take note of empirical evidence, but simply that they usually took it for granted that it should be interpreted in the light of Christian doctrine; there was little biblical literalism—what Coleridge later called bibliolatry—in this period. In France, there was more polarisation; and Buffon was for example censured by the Doctors of the Sorbonne for some of his wide-ranging and quasi-evolutionary writings.

There was no dramatic change as geologists came to accept that only evidence from rocks should count in their science. Hutton, who was the first to argue for a strict principle of uniformity in geological explanation, did so in the course of a paper (later expanded into a book) which is full of natural theology. His paper was read in Edinburgh in 1785; its concluding words, 'we find no vestige of a beginning,—no prospect of an end,' shocked some contemporaries, but really represent limits imposed upon geological speculation rather than an assault upon religion. In the next generation, men of science found themselves apparently forced by their belief in the general uniformity of nature to posit catastrophes in the past. How for example could one account for the entire mammoths found quick-frozen in the wastes of Siberia without supposing that the climate had suddenly changed? William Buckland, Professor of Geology at Oxford and subsequently Dean of Westminster, wrote in 1823 a book full of evidence for Noah's Flood; in the course of this research, he had given bones to living hyenas to confirm that they crunched them in the way bones found in a cave in Yorkshire had been crunched, and that their faeces had the same character as those found there. This disagreeable programme indicates a firm belief in the uniformity of nature over a long period of time. What did constrain men such as Buckland and Cuvier was the belief that in Genesis one could find a time scale for the earth's history, which began in or about 4004 BC.

In 1830 Lyell published his book which slowly convinced

geologists that past changes were all to be accounted for in terms of forces operating today; which in turn entailed a very much longer history for the earth. Some geologists felt it necessary to expand the 'days' of the account in Genesis to mean millions of years each; but the majority were not as literal minded as this. Then a generation later came Darwin's theory; though the impact of this outside biology has probably been exaggerated. The anonymous bestseller *Vestiges*, which came out in 1844, had brought the idea of evolution before the public mind; and the vision of a world on which man was a latecomer, and on which species had come into being and become extinct following some inexorable law, was becoming familiar. What Darwin did was to convince men of science of evolution, by argument based upon numerous examples; so that evolutionary theory could no longer be treated as mere speculation.

We are apt to suppose that all religious people were thrown into ferment by the publication of Darwin's theory; but this seems not to have been true. Much more disturbing was the so-called higher criticism, or biblical criticism; in which the Bible was analysed like any other historic document or collection of documents. Critics noticed that the first two chapters of Genesis contained two incompatible accounts of the creation; and in general they examined the various books of the Bible, seeking to separate fact from interpretation or interpolation, just as one would study an old history book, or a version of a speech said to have been made by some statesman. This had been going on for some time, particularly in Germany; but it was introduced into Britain in *Essays and Reviews,* a book which was published in 1860 and aroused a storm. There was little science in it, though one of the essays was against belief in miracles and another showed the impossibility of reconciling geology and Genesis.

By about 1880 most scientists and most theologians had come to accept the theory of evolution; but it is worth remembering that this was also the great period of foreign missions, and that for every clergyman or college tutor who painfully lost his faith —and the process seems usually to have been painful—there were several people who went off to be missionaries. In the same way, it would be a mistake to suppose that political thinkers were all converted to social Darwinism, and believed that the weak must inexorably go to the wall; for this period saw the rise of socialism, and indeed A. R. Wallace, the co-discoverer of the principle of evolution by natural selection, was a socialist and advocated nationalisation of the land. We should not over-estimate the effect on the majority of people that scientific discoveries have; though in the present case one must no doubt admit that Darwinian rhetoric was of assistance to secularists and to *laissez-faire* politicians.

Theology and science have had these close relations partly because, down into the middle of the nineteenth century, many scientists were clergymen, and most were religious men; and more seriously because both scientists and theologians have often or usually seen it as an important part of their task to explain what goes on in the world. Some men of science have rejected this view of science; preferring a positivistic account, in which science consists of laws linking observables, and therefore de-scribes concisely rather than explaining. In the same way, some theologians have urged that religion is a matter of coming to terms with whatever happens, and seeing Providence in it; rather than of explaining anything, for imperfect men could never comprehend the inscrutable purposes of God. If either party adopts this view, then there can be no conflict; but few creative scientists can really have adhered to this view of science, and few religious men have believed God's ways to be com-pletely unlike our ways. So we have had this history of inter-actions and tensions; and while there have been conflicts—and one should remember that critics of Galileo or Darwin often had a good eye for a poor argument—it is probably true that both science and theology have been gainers from this interplay, and that for most of their history the two have been com-plementary.

The sciences have also had close relations with the fine arts. Music has since the time of Pythagoras been closely linked to mathematics; and many of the great scientists of the seventeenth and eighteenth centuries concerned themselves with harmonics. Galileo's father was a distinguished musician; and it is ironical that it was Kepler rather than Galileo who was fascinated by the idea of the harmony of the spheres, and even wrote down the notes appropriate to the various planets. Natural philo-sophers made experiments on frequencies and scales; and in the

mid-nineteenth century Helmholtz tried to produce a physiological explanation of how we delight in harmonies and reject
discords.

With the visual arts the relationship has been probably even
closer; for not only has optics always been a leading branch of
physics, but natural history has always depended upon good
illustrations, and conversely portraiture and sculpture have required some knowledge of anatomy. The most direct contribution of physical science to drawing and painting has been
probably the theory of perspective; which again goes back to
antiquity, but became prominent in the Renaissance and in the
seventeenth century when pictures giving the illusion of windows
were popular. Painters actually worked with a device called a
camera obscura, or dark little room, in which an arrangement
of mirrors threw an image of the view outside upon a table.
The image could then be drawn round. Portable versions of this
device were invented; a famous kind being the *camera lucida*
of the crystallographer W. H. Wollaston, invented at the beginning of the nineteenth century. Wollaston is also famous for his
paper to the Royal Society describing work done with the portrait painter Sir Thomas Lawrence on why the eyes in portraits
seem to follow one around.

The *camera obscura* and its portable descendents were valuable particularly to those who had to draw landscapes but had
only a moderate talent for it. Drawing was in the eighteenth and
early nineteenth centuries a necessary accomplishment not only
for well brought-up ladies but also for army officers, sea captains,
surveyors, geologists, and explorers, whose description of a piece
of country was much improved by illustrations. The India
Office Library in London, for example, contains a fine collection of drawings by amateurs in this period. The next stage of
development of the *camera obscura* rendered this skill less
necessary; this was of course the invention of photography,
where the image falling on the plate is recorded by chemical
means and no longer needs to be traced around by the pencil.

Such drawing was clearly not a highly creative or interpre-
tive process; and mathematical theories of perspective rigorously
applied give a cold perfection or sometimes an odd effect in a
picture. Goethe in his *Farbenlehre* of 1810 pointed to the in-
adequacy of the existing 'Newtonian' theory of optics and

perspective to account for what we actually see. People do not
seem to shrink as they go away from us as the theory of perspective would imply; our judgements of size and distance are
more complicated, and colours in particular give depth to a
picture as effectively as does perspective. Goethe was fascinated
by those pictures of cubes which seem either to be coming
forward out of the page or receding into it, depending on how
one looks at them; and by the way in which a black circle on
a white ground seems smaller than a white one on black. His
book failed to shake the faith of most physicists in the received
mathematical theory of optics; they were engaged over the next
twenty years in the less general debate over whether light could
be best represented as waves or particles. Painters were interested
by this contribution to what would now be called psychology,
and when Goethe's book was published in English, it was translated by Charles Eastlake who was both a Royal Academician
and a Fellow of the Royal Society.

One route from science into drawing therefore led towards
a rather mechanical method of doing landscapes, and ultimately
to photography; while another, based upon a criticism of a
narrow and abstract theory of optics, led in the end towards
new experiments with shapes and colours when in the later
nineteenth century it became clear that the photograph could
replace a mechanical kind of drawing, and that geometrical
perspective was only half the story. Science was also involved
in the actual pigments used in painting. Here as often, the
process seems to have begun with chemists analysing pigments
(Davy for example investigated those used in Herculaneum),
and then trying to improve upon them. The synthetic dyes of
the late nineteenth century opened new possibilities for the
painter as for the fabric designer.

Men of science also concerned themselves with the problem
of classifying colours; the astronomer Tobias Mayer was among
the first to try to do this, in the mid-eighteenth century; he was
followed by Werner the mineralogist at the end of the century,
who hoped to be able to describe colours unambiguously and
exactly so that they could be incorporated into descriptions of
crystals, and whose book was translated into English by the
great flower-painter James Sowerby; and by the chemist
Wilhelm Ostwald at the end of the nineteenth century—

Ostwald's system forms the basis of those now in use. The problem is that colours cannot be plotted on some single scale like musical notes; a three-dimensional chart is required, taking note of what are now called hue, brightness or value, and saturation or chroma. As a tailpiece to any discussion of colour, we might add that it was the chemist John Dalton who found that he was colour-blind and published the first scientific account of colour-blindness in 1798. As often happens, the study of why some people cannot see colours has helped in understanding how most people do see them.

Just as any form of survey had to be supplemented by topographical sketches, because often a drawing can tell us more than a paragraph; so natural history and anatomy have always required illustration. This need has even now not been superceded by photography. An artist can without actual distortion emphasise those aspects of his subject which are important, whether he is drawing a specimen or a scene; botanical art differs from drawing flowers, for example, in that there is an agreed list of characters which must be shown, and that this requirement must not be subordinated to the need to be decorative. Again a photographer must take a particular plant or animal, or a landscape at a particular time; while the artist can generalise. In natural history, he is not simply painting the individual, but the species; if he is the first to do so, his picture may become the 'type' against which specimens may be compared to see whether they belong to the same species as that which he has depicted. He must therefore not draw attention to individual peculiarities; and his task will be much easier if he has several specimens to work from.

The great artists of the Renaissance produced splendid anatomical and natural history drawings, in which the knowledge displayed of anatomy or botany often goes well beyond that in the texts being illustrated, or current at the time. With the coming of printing, it became possible to get these illustrations reproduced in quantity; manuscript reproduction is fatal to natural history illustration, which each time it is copied becomes more stylised and further from the original. At first woodcuts were used to print illustrations; these were often cut by the artist who had drawn the picture, and also had the advantage that they were so made that what was to print black stood up from the wood in relief, like type does, so that they could be printed with type. But in the seventeenth and eighteenth century, they were largely displaced by copperplate illustrations, where the picture was engraved (or sometimes etched with acid) on a plate of copper, so that finer detail could be shown. This is an intaglio process—that is, what is to print black is below the surface of the plate; it must therefore be printed separately from the text, and illustrations no longer appeared on the page to which they referred but were separately inserted. A further problem was that as a rule the artist could not be his own engraver, so that a craftsman stood between the man who drew the picture and the reader who saw the printed version.

At the beginning of the nineteenth century, science came to the rescue of the artist with the invention by Senefelder of lithography, which depends upon the repulsion between oil and water. A picture is drawn upon a prepared stone with a greasy medium; the stone is then wetted and inked, and the greasy ink will adhere only where the drawing was done, and so prints can be made from it. Drawing on stone was much easier and cheaper than engraving on copper; and for fine illustrated books lithography had by the mid-nineteenth century become the accepted medium. It was not until the latter part of that century that full colour printing became a practical proposition; and until well into the twentieth century the finest coloured illustrations were done by hand, the outlines being printed by lithography or copper plate. In recent years there have been some splendid printings of plant and animal paintings of the eighteenth and nineteenth centuries done from the original drawings; so that we have in some cases more faithful reproductions than our forefathers had.

Contemporary with the invention of lithography was the introduction of wood-engraving, in which the greatest figure was Thomas Bewick. This is done on the end-grain of boxwood, and is both very strong and can show great detail even in small pictures. It can, unlike lithographs or copper plates, be printed with type; and therefore brought the illustrations back onto the printed page in reasonably priced works of science. Bewick became a respectable natural historian though again his plates are much better than his text. He did some engravings for the *Family Herbal* of Robert Thornton; who is chiefly remembered for his magnificent *Temple of Flora*, with its grandiose and theatrical illustrations. In fact in this great book botanical value has been sacrificed to romantic excitement; Thornton and his artists did not achieve the balance of the accurate and the dramatic which is found in Audubon's bird pictures. In an earlier work, Thornton had employed William Blake as an engraver; but made an enemy of him when he saw 'less of art than of genius' in the plates Blake produced.

Blake had done a plate for Erasmus Darwin's *Botanic Garden* of 1789–91, which is a long poem mostly about the Linnean system and the sex life of plants but including in text and notes references to most of the science of the day. From the very beginnings of modern science the connections between it and literature have been strong; though sometimes, as with Blake and Keats, there have been elements of hostility in the relationship. One of the first things that the Royal Society did in the 1660s was to set up a committee to purify the language of science by bringing it closer to that of artisans, who were believed to call a spade a spade in contrast to academics who were thought to go in for elaborate circumlocutions. John Dryden was one of those who served on this committee. Certainly the language of poetry, of sermons, and of essays became much simpler in the later seventeenth century; and it seems likely that the demand of natural philosophers for a plain expository prose, aiming at clarity rather than metaphor or paradox, had much to do with this development. This restriction to plain, and sometimes flat, language was something against which those associated with the romantic movement were to protest, about the end of the eighteenth century.

Not only the style but also the subject matter of literature was affected by science. Science fiction goes back at least to the early seventeenth century; an early example being the *Somnus* of Kepler, describing a trip to the moon in a dream. The point of this little work was to show that dwellers on the moon would imagine that the earth and all the other heavenly bodies went around them, and therefore that wherever one was in the universe one would assume oneself to be at the centre; and thus to weaken the traditional arguments against the Copernican system, in which the earth was no longer at the centre of things. In the course of the seventeenth century, a whole series of works were written on the plurality of worlds; arguing that if the earth were a planet, then other planets must be like it, and therefore will have their inhabitants too.

The most famous author in the genre was Fontenelle, who thereby popularised the Copernican system and the mechanism of vortices proposed by Descartes. His entertaining little book, first published in 1686, was largely responsible for getting him the appointment of Permanent Secretary to the Academy of Sciences; in which capacity he wrote obituaries of many great men of science, including Newton. Fontenelle had his Jovians phlegmatic, as befitted dwellers on a cool planet; while his Venusians were amorous and negroid. The great astronomer Christiaan Huygens extended living beings to the hypothetical planets encircling distant fixed stars; for natural philosophers were still sufficiently teleologically minded to be puzzled about the purpose behind the stars. Many fewer would have done for navigation; while of the most distant it was absurd to say that they had been made merely to be viewed through our telescopes. They must therefore be suns which illuminate planets upon which there are rational beings—for why should our sun be the only star thus dignified?

These works are perhaps closer to popular science than to science fiction as most people might understand the term. But science also came in for satires, of which the most famous was probably *Gulliver's Voyage to Laputa,* a detailed joke at the expense of contemporary natural philosophers—which fact perhaps makes it the least readable of the Gulliver stories today. Swift and Fontenelle were also involved, on opposite sides, in the literary battle between the Ancients and the Moderns, which was a kind of hundred-years war beginning about the 1620s. The Ancients argued, in the spirit of the Renaissance, that we were but dwarfs on the shoulders of giants, and could not expect to surpass the achievements of antiquity. The Moderns, for whom Fontenelle became a spokesman, said on the contrary that in many ways antiquity was already being surpassed; that if in the realm of poetry, drama, and rhetoric the question was uncertain, it could not be doubted that in all the sciences the men of the later seventeenth century outdid the Greeks and Romans.

The question could not be easily given a complete answer;

but in Britain the Ancients received a heavy blow from Richard Bentley, the leading philologist of his day; for they had held to the doctrine that the older a work was, the purer its style must be, and had greatly admired the letters of Phalaris—which Bentley in 1697–99 proved from internal evidence to be late forgeries. Bentley was also famous for his confutation of atheism in a series of lectures which included the first attempt to popularise Newton's work on gravitation. Nearly a hundred years before Bentley's demolition of the letters of Phalaris, a similar critique had weakened the hold of magic; for in 1614 Isaac Casaubon showed that the writings attributed to Hermes Trismegistus, who was supposed to be only slightly later than Moses and therefore only slightly less authoritative, dated in fact from the Christian era.

Discoveries with the microscope and telescope had a profound effect upon the imagination of literary men in the eighteenth century, when the age of two cultures was still far off. Such classic and representative works as Thomson's *Seasons* and Young's *Night Thoughts* (as well as Erasmus Darwin's curious verses) are full of science; as later was Tennyson's *In Memoriam*. Coleridge went to the chemistry lectures of his friend Davy to increase his stock of metaphors and to get arguments against materialism; Goethe derived inspiration from both alchemy and modern science; and Darwinian theory provided a world view that writers could like or hate. Certainly any historian of literature must have a general acquaintance with the history of science.

In the twentieth century the situation may have become slightly different. Most men of science have in our century been in some sense professional scientists. They have earned their living by their science; but more importantly they have been educated differently from historians or linguists, and speak in some sense a different language. Down to the 1860s it was possible for the layman to read and understand a much higher proportion of original publications in science than is the case nowadays. Men of science had to write for a general readership even in scientific journals, and many did write with more style than is now generally met with. Nowadays laymen have to rely upon popularisers; and to popularise, as to translate, is always to some extent to betray the original.

But this is not the whole story. To make much of most work on astronomy in the early nineteenth century one needed a proper training in mathematics; and for anatomical papers, a medical training; although the layman could be excited about the discovery of new planets, or of strange creatures like the duck-billed platypus. But there were some sciences in a state of ferment, which had not yet acquired a forbidding technical language and theoretical structure, or in which these had been overthrown in a revolution and not yet replaced. Thus it was to lectures on electricity that people flocked, to see electric-shock machines demonstrated and lightning simulated, and to hear speculation about the connection between the electric and the nervous fluids. Similarly, they went to lectures on geology, where they heard of the economic importance of coal and of canals (and therefore the value of geology), heard how geological evidence supported (or occasionally, in some secularist halls, confuted) the biblical account of Noah's Flood, and heard about the extraordinary creatures which had inhabited the earth in remote ages. In chemical lectures, they heard in the same way about the bleaching and tanning industries, and about gun-powder which would prevent modern civilisation being overrun by a set of barbarians like the Goths and Vandals; and they saw dramatic demonstrations of chemical reactions. They heard how little chemical theory there yet was; and were treated to speculations about the connection between electricity and chemical affinity.

By the middle of the century, electricity was becoming a mathematical discipline, though Faraday could still get across some of the old excitement and generality. Similarly, geology had become much more technical, and fossils were no longer extraordinary productions but something to be counted and assessed in dating formations. Chemistry had made the most rapid progress towards a forbidding technicality; and it would have been very difficult to get a lay audience to feel that they were on the frontier of knowledge, as Davy in London and Fourcroy in Paris had done half a century earlier. The amateur could no longer feel involved and on a level with the pro-fessional.

This was something which had happened much earlier in astronomy. Copernicus had written for mathematicians; but

Galileo with his telescope had brought in the layman, and Descartes' account of planetary motion was both splendidly speculative and readily comprehensible. Newton's work was on the other hand very difficult both because it was mathematical, and because the conception of universal attraction across void space was not easy to grasp. Newton therefore demanded popularisers, of whom Voltaire is probably the most famous. Even in natural history, the change from the Linnean system to the natural system in the early nineteenth century meant that the amateur could probably no longer have the knowledge to classify specimens. Most of the sciences most of the time belong to the sphere of the expert; and only at periods of rapid change, or in their earliest years, are they completely accessible to laymen.

Lay interest in speculation, or rather perhaps in cosmology in the widest sense, has meant that often the least 'respectable' sciences at any time have been, and still are, the most popular ones. Popular attention, that is, is focussed more often than not upon the fringe of science; and it is from this fringe that writers borrow their imagery. These fringes have sometimes developed into part of the established corpus of science; but sometimes they have not, and can with the wisdom of hindsight be called 'pseudo-sciences'. In any event, it is naturally enough very rare for all the speculations of a pioneer to turn out to be well-founded.

In the mid-eighteenth century, mechanistic theories of physiology and psychology were all the rage; men's bodies were compared to machines, and slightly more elaborate models of their minds were proposed. David Hartley took up a suggestion from Newton and Locke, and urged that ideas were vibrations in the brain, and that various vibrations became associated together giving rise to chains of thought. It was hard to see how human free will had a place in this scheme, but determinism was not something to be feared in Hartley's day. On the contrary, the universal reign of law rather than of caprice seemed to indicate the possibility of progress. But by the end of the century, physiologists could show the implausibility of these mechanistic accounts of bodies and minds, for the workings of both were too complex to be described in such crude terms. Philosophers were also prepared to argue that statements about minds could not be reduced to statements about brains; that the two things were in different categories.

This is a debate that has gone on for a very long time indeed, and no empirical discovery makes much difference to it. In the early nineteenth century, general attention wandered from the technical writings of neuro-physiologists to the exciting and easy science of phrenology, or the study of bumps on the head as a clue to the shape of the brain, and hence to the cast of the mind, of the person being studied. This science developed out of physiognomy, the science of determining people's character from their faces; and turned out to be equally weak as a guide. It was unfortunate when a notorious murderer turned out to have the 'bump of benevolence' particularly strongly developed; but the science survived even this apparent falsification, just as astrology had survived when predictions had gone wrong. Phrenology was especially popular with some educationalists, notably George Combe of Edinburgh, to whom it seemed to promise a means of gauging the ability of children, rather like IQ tests in the twentieth century. On one level, phrenology could lead to amusing parlour games; while on another it seemed to promise a prospect of understanding and predicting human actions, and of bringing people's behaviour within the realm of universal natural law. Phrenology can also be seen as a groping towards theories of cerebral location; but here there are great dangers of anachronism, and certainly it was not for this that audiences flocked to phrenology lectures.

As towards the middle of the century interest in phrenology began to wane, there came a new excitement as spiritualism was introduced to Europe from America. In the seventeenth century there had been great interest in ghosts and in witchcraft; and some among the early Fellows of the Royal Society were firm believers in the occult. Joseph Glanvil, for example, who was one of the leading propagandists for the new science in general and for the Royal Society in particular, was also the author of a collection of ghost stories in which he believed, and indeed argued that scepticism about ghosts, devils, and witches would soon lead to religious scepticism generally. But by the end of the seventeenth century few men of science would have thought that; they believed in a mechanical world, with its evidence of a divine designer.

With the coming of the Darwinian revolution, the argument from design seemed to be weakened; and a number of eminent men at the end of the nineteenth century founded their faith in human immortality upon the evidence of spiritualism. Men of science were soon involved in investigating what was happening; and Faraday and Tyndall were among those who found their scepticism confirmed and who detected deliberate or unconscious fraud. Others were uncertain, because while some effects could be accounted for their seemed to be a core of genuine phenomena. Henry Sidgwick, who taught moral philosophy at Cambridge, was the most important figure in the Society for Psychical Research; which investigated hauntings and seances, and published accounts of them. In 1886 it published a large book, *Phantasms of the Living*, which includes a number of experiments on telepathy as well as many stories of those who have seen an apparition of someone else, who often turns out to have been in a crisis or dying at the time. The book is fascinating for its sceptical handling of dubious data; and is interesting for the list of distinguished members and patrons of the Society at the front. As well as Mr Gladstone, Mr A. J. Balfour, Lord Tennyson, and John Ruskin, this list includes J. C. Adams the discoverer of the planet Neptune, J. J. Thomson, Lord Rayleigh, Oliver Lodge, William Crookes, and Balfour Stewart among physical scientists, and A. R. Wallace, the co-discoverer of natural selection, among biologists.

The Society's investigations appeared inconclusive, and popular attention waned; and it was the psychological theory of Freud which next aroused enthusiasm, and affected the way most people looked at human relationships—though only parts of it found their way into the mainstream of psychology and anthropology. Meanwhile there had been developments of the Darwinian world view; by the end of the nineteenth century Malthusians, who had through the century been trying to persuade the working classes to have smaller families, turned to the gloomy science of eugenics and worried that the finest lines of inheritance would be swamped. Others studied animal behaviour, first for its own sake—and the invention of binoculars made this possible for ornithologists, for example—but then for the supposed light it might cast upon human behaviour; and in our own day those following Darwin in the study of the interactions between different organisms sometimes provide support for an apocalyptic literature.

In all these cases one has analogies being taken too far, and elaborate structures of hypothesis being erected on slender foundations; but after all some of the most established sciences have grown out of just such confused matter. What the history of fringe science surely shows is the need to be sceptical about science; scientists need models, metaphors, and analogies and their statements need not always be seen as sober statements of facts. On the other side, scientists have often felt gloomy about what seems to them the irrationality of the world in general, whether they are contemplating phrenology, spiritualism, or the so-called anti-science of today. Science is only in some senses cumulative. We should not suppose that we are wiser than our ancestors, but we need not believe that we are sillier. Men of science, and particularly philosophers of science, have sometimes put forward the scientific community as an arbiter to separate real from false knowledge. Whether or not we believe that, the emergence of such a community is an important matter, and one to which we turn in the next chapter.

4

THE SCIENTIFIC COMMUNITY

Those who, like Derek de Solla Price, count the people involved in science, or the books and papers written on scientific subjects, have pointed out that both these have increased very rapidly since the seventeenth century. Indeed the increase has been exponential, so that the numbers involved have kept on doubling every few years. Unless the rate of growth eases off, as it seems to be doing, there would about the end of the twentieth century be as many scientists as people, and they would have no hope of being able to read more than a tiny fraction of the papers and books published in their field.

We need not take such predictions too seriously, and we must be very wary of classifying men as 'scientists' and activities as 'science' before the nineteenth century when these terms came into use in something like their present sense; but it is clear that since the middle of the seventeenth century what we can call the scientific community has come into being and has grown rapidly. As science has won its autonomy from other intellectual activities such as philosophy and theology, so its practitioners have emerged from other academic fields, from professions such as medicine, and from practical activities such as instrument-making, and have become conscious of themselves as scientists.

The growth in the scientific community has probably not been accompanied by a corresponding growth in the number of men of genius involved in science. While men such as Boyle and Newton were not typical of the early members of the Royal Society, for example, there do seem to have been in the seventeenth century an extraordinarily high proportion of powerful and original thinkers and experimenters involved in science. Partly this was no doubt because there was less to learn before one reached the frontier of knowledge, and therefore more scope for originality; and partly because the excitement of the new philosophy, as it was called, attracted the most able men of the day into science. But partly this state of affairs seems to indicate the lack of a scientific community. In antiquity, there had been great men such as Aristotle, Archimedes, and Ptolemy; but in a period extending over several hundred years (Ptolemy's work was done about five hundred years after Aristotle's, and these two do not represent the beginning and the end of science in the ancient world) we find such great men appearing from time to time, but apparently without the steady support from dozens of competent people carrying the work steadily forward between times. In the medieval period there was again no dearth of science, as ancient texts were translated and studied and occasional original thinkers went beyond them; but again there was no steady advance of knowledge such as we have come to expect in the last three hundred years.

One reason for this lack of steady progress was that in the days before printing, information no doubt circulated more slowly and less accurately than it did later. But with the invention of printing, and the Renaissance of the fifteenth century, there did not come an upsurge in scientific activity; indeed, at first the emphasis upon good texts and accurate translations went with a veneration for the writings of antiquity that was inimical to experimental science. As it gradually became evident that the great men of the past had not always agreed with one another, and had sometimes made what were clearly mistakes of fact or interpretation, there came in the sixteenth century a sceptical reaction; and experimental science came to seem a humbler but safer way of attaining some certainty in knowledge.

In Renaissance Italy there had been various academies, loosely modelled upon Plato's Academy in Athens. These were small groups of men interested in intellectual matters; they were as a rule chiefly literary, but one in Naples, of which della Porta was the leading light, had in the mid-sixteenth century concerned itself with natural magic, which included much that we would classify as the science of the day. An account of the researches of these *otiosi* was published in 1558, and put into English a hundred years later; it includes recipes for beautifying women as well as experiments on magnetism and the first description in a European language of a plane surface kite. Porta lived on until 1615, and in his old age was Vice-President of the Academy of the Lynxes, which was under the patronage of Prince Cesi and whose most prominent member was Galileo. This academy was dissolved at the time of Galileo's troubles with the Church; but his disciples formed another academy, the *Accademia del Cimento* (Academy of Experiments) patronised by Prince Leopold of Tuscany; their proceedings were published as the society came to the end of its ten years of life in 1667, and were published in an English translation in 1684.

At much the same time, there were academies, societies, and groups of people interested in science forming in France and in England. In France, the Minim Friar Mersenne was the most important figure in the organisation of science; and his correspondence in the middle of the seventeenth century enabled natural philosophers to keep abreast of what was going on. Anybody making a discovery would pass the news on to Mersenne, who would then inform his numerous other correspondents. Such informal methods of circulating information lasted a long time; thus Sir Joseph Banks who was President of the Royal Society from 1778 to 1820 probably wrote about fifty letters a week during his active life; and in our day, too, the correspondence of men of science is important and reveals much that does not appear in printed sources. But in the 1660s came an innovation that made correspondence or actual meeting no longer the only way of keeping up with developments; for the scientific societies by then established in Paris and in London began to publish, or rather to have published under their aegis, scientific journals.

These societies are particularly important because they have survived down to the present day, though the Parisian one has undergone various changes of name and constitution. The London society was the Royal Society, whose object was the improving of natural knowledge, and which received its first royal charter in 1662. Many of the leading members of the society had belonged to earlier informal groups, associated with Gresham College in London where public lectures were given on a wide range of topics, and with Wadham College in Oxford, where John Wilkins was Warden during the Commonwealth and attracted many able men, including Christopher Wren and Robert Boyle. Wilkins was Oliver Cromwell's brother-in-law; but he made his peace at the Restoration, and became one of the two secretaries of the Royal Society and later Bishop of Chester. Francis Bacon had in the 1620s described a Utopia, or *New Atlantis*, where the most important institution was Salomon's House, a place of learning where salaried men of science pursued their investigations collectively; and in the 1650s there had been various plans in circulation for such an institution in England. But in the event, the restored King Charles II, while sympathetic, could not afford to subsidise the Royal Society; which became a body in which a large membership supported by their dues the researches of a minority of active members.

In France things were ordered differently. The Academy of Sciences was inaugurated, soon after the Royal Society, as a Department of State; the Academicians received a small salary, and their numbers were limited. In return for this state support, they had to undertake researches on practical matters, such as the improvement of street-lighting or of gunpowder, when ordered to do so. Down to the twentieth century, it was possible for enemy aliens to belong to the Royal Society; but the Academy of Sciences being an official body took a different view, and Christiaan Huygens, one of the leading early members, had to resign when war broke out between France and Holland, although the Anglo-Dutch wars had not affected his relations with the Royal Society. Henry Oldenburg, one of the Secretaries of the Royal Society, was however suspected of spying at the time of these wars, largely because his foreign correspondence was so extensive; but after a short time in the Tower of London, he was released.

Both these societies sponsored, though they did not at first officially publish, scientific journals. The French one was published in Paris and Amsterdam, and was called the *Journal des Scavans*; it concentrated upon reviews, and thus kept its readers abreast of developments in a wide field. In the sciences, as in the arts, there are still journals which perform this important function; assessing the work of numerous people in a field, and making it known to others who might otherwise have been in ignorance. As the number of those engaged in science grew, and it could no longer be expected that everybody would know everybody else, this reviewing function of journals became

85

essential. Despite the existence of reviewing journals, one often comes across cases in the past of men of science who were unaware of the relevant work even of eminent contemporaries; and things are no doubt not very different today.

One way of investigating the scientific community is to see what names are cited in reviews; but while reviews are important, they do not themselves usually provide a vehicle for the announcement of discoveries, though they may offer an opportunity for the reviewer to propose some new interpretation which is just as important as the discovery of new facts. The reports of the Italian academies had described the discoveries made by the members; but they were collective reports, which only appeared after the academies had ceased work and therefore have a retrospective character. It was the *Philosophical Transactions,* first published by Oldenburg as a speculation, continued by later Secretaries of the Royal Society, and only an official publication since the mid-eighteenth century, which was the first journal to publish signed original papers. This pattern has since become the norm for scientific journals.

With the coming of these various journals, it became much easier for those whose talents were less than those of a Galileo or a Harvey to assist in the progress of science. A discovery communicated to the editor of a journal was quickly circulated in print around the scientific world; and papers could be short, so that whereas earlier a man waited until he had enough material to fill a book, now he could publish an account of a single experiment soon after it was performed. Science has always been in principle public knowledge; and with the coming of journals this ideal could be approached more closely, and the pace of scientific advance also quickened. The contributors and subscribers to the journals formed a scientific community, though naturally some were much more committed to the increasing of natural knowledge than others.

These journals brought into being what has been called 'normal science', concerned with questions difficult to answer in contrast to 'revolutionary science', concerned with questions difficult to ask. Papers of greater or less interest were submitted to the editor, who passed them on to some expert who acted as referee, recommending publication perhaps, or else rewriting, or maybe rejection. The referee was supposed to repeat the experiments described; and publication of experimental researches in journals with this element of scrutiny, meant that it became generally accepted that experiments reported should genuinely have been performed, and be sufficiently clearly described to be easily repeatable. This had not always been the case in the past; and Robert Boyle is supposed to be one of the first men of science to publish uncooked experimental data. His predecessors had not been dishonest, for this was not seen as a question of ethics before the 1660s; they had simply not bothered to perform an experiment whose outcome seemed obvious, or had averaged out results omitting observations that seemed poor.

Science in journals was not only public knowledge; it was normal science in that papers usually represented a relatively small advance along a tolerably well-marked road. Chemists analysed a mineral water and physicians described an interesting case in a manner that was accepted; and the amount of information available thus steadily grew, for this normal science is cumulative. When the rules gradually lead to anomalies, or the general picture is lost in a mass of detail, then a revolutionary scientist may come along and suggest some quite different way of looking at the phenomena; but to do so will require more space than a journal can provide, and such new suggestions have usually been most effective when presented in books.

The editors of journals, and the referees they consult, thus played and play an important part in the propagation of the approved approach to problems, methodology, and terminology in science. Often the names of their contributors, who may have discovered something dramatic or invented something useful, are much more familiar to us; but if we are to see science as a social activity in which the advancement of knowledge cannot be separated from the dissemination of knowledge among scientists, then we must look hard at editors and secretaries of societies. Insofar as there was a community of men of science, it was these people who formed and guided it, and gave it its identity.

We saw that the Dutchman Christiaan Huygens was a prominent member of the Royal Society and of the Parisian Academy of Sciences; the scientific community was an international one. But communication across national and linguistic frontiers was imperfect. The seventeenth century saw the rise of the vernacular languages in Europe; so that while at the end of the sixteenth century academic works had been written in Latin, and then popularised or perhaps translated into English or French, later seventeenth-century science was published as a rule in a modern language. Galileo published his comparison of the Copernican and Ptolemaic systems in Tuscan, which had been a dialect but was becoming the written Italian language; and indeed it seems that publishing in the vernacular rather than in the language of scholarship, and thus making his arguments available to anyone who could read, was a part of his offence in the eyes of his accusers. The journals of the scientific societies were for the most part in French and in English; but in Germany Latin survived longer as the learned language.

Many men of science could read modern languages other than their own; but foreign periodicals are always less easy to come by, and even for those who have learnt a foreign language it is not easy to respond to the more subtle and theoretical parts of a paper or book. Translation from one language to another therefore became necessary if science carried on in one country were to be fully understood in another. It is not surprising that in the seventeenth century there was great interest among men of science in the possibility of a universal language, in which the symbols would stand for things or concepts (like the 'Arabic' numerals do) rather than for sounds. The model for these was the Chinese characters; for the Jesuit missionaries had reported that Chinese who could not understand each other's spoken dialects could communicate by writing. Wilkins proposed the most worked-out universal language in his *Essay towards a Real Character* of 1668; but like later proposals for artificial universal languages, it never came into general use, and the scientific community was divided by barriers of language. Down to the end of the seventeenth century, really technical works with a prospect of small sales in any one country still appeared in Latin; the prime example being Newton's *Principia* of 1687, of which the English translation did not appear until 1729, soon after Newton's death.

After the journals had been coming out for some years, it became difficult to find quickly all the papers on some topic; even given the indexes which came out from time to time. Various abstracts and selections from the journals were accordingly made. Thus the *Miscellanea Curiosa* of the early eighteenth century reprinted papers from the Royal Society's *Philosophical Transactions*; and at the beginning of the nineteenth century came an enterprise of greater extent, the publication in nineteen large quarto volumes of an edited version of the complete run of the *Philosophical Transactions* from 1665 to 1800. Some of the papers are reprinted complete; of others only the title survives; many are abridged; and the work had an index and brief biographies of eminent contributors. There were various selections in translation from the works of the French *Virtuosi* of the Academy of Sciences published in English in the late seventeenth century; and in 1742 an ambitious translation and abridgement of the *History and Memoirs* of the Academy appeared, undertaken by John Martyn a botanist and Ephraim Chambers the compiler of an early encyclopedia.

The first journals were thus published in association with the scientific societies, and anybody wanting to keep up with science would have to read them. And indeed the early volumes of the *Philosophical Transactions*, for example, are not, and were not, daunting to the general reader; though there are some mathematical and astronomical papers which many Fellows of the infant society must have skipped through. But in general there were not enough experts for anybody to direct his papers to experts alone; and with the title of Natural Philosopher, generally given to men of science down to about 1850, went an expectation that science was more than a collection of facts, and that its devotees should form and propagate a world picture.

The scientific community then, in the first century of the existence of the scientific societies, was not a fragmented body, although there were linguistic barriers between men of science in different countries. Within the scientific community there were naturally divisions, between the natural historians, the mathematicians, and the chemical philosophers for example; and these divisions sometimes became apparent, for example in the election of a President of the Royal Society. There were also social barriers, for although the Royal Society surprised some foreign observers with its relative informality and democracy in the seventeenth century, its membership was even then chiefly made up of gentry and professional men; when a tradesman, John Graunt the statistician, was admitted this seemed to the society a matter for self-congratulation. In the eighteenth century, the fashionable character of the Royal Society did not diminish; and it was socially out of tune with the rising self-made men who were involved in the Industrial Revolution.

These different worlds can be seen exemplified in the careers of two eminent contemporary chemists, Humphry Davy and John Dalton, who grew up in the late eighteenth century. Both came from a relatively humble background far from any great city. Davy was apprenticed to an apothecary in Penzance, and his intelligence attracted the attention of some local gentry; then the son of James Watt was sent to board with Davy's family in the hope that the mild climate of Cornwall would relieve his consumption, and he and Davy became friends. Some experiments done with improvised apparatus by Davy were brought to the attention of Watt's friend Thomas Beddoes, who had taught chemistry at Oxford and had like Watt belonged to the Lunar Society of Birmingham, a society which had brought together men of science, industrialists and doctors. Beddoes appointed Davy as his assistant at an institution he was setting up in Bristol, with funds provided by Josiah Wedgwood the potter, in the hope of curing diseases by the administration of gases. Davy discovered the properties of laughing gas; and on the strength of these and his other researches was appointed to a lectureship at the newly-founded Royal Institution in London.

He now left the world of the provinces behind him; for the Royal Institution was supported by the fashionable world, and particularly by landowners anxious to improve their stock and their crop yields with the help of science. Davy duly lectured on fertilisers and on tanning, and also on chemistry generally; and his lectures were immensely successful, attracted large audiences, and put the Royal Institution on a sound footing. He meanwhile did fundamental work on the relations of electricity to chemistry, on the nature of acids, and then on the reasons for the explosions in coal mines. He was knighted and then made a baronet; was elected Secretary, and ultimately President, of the Royal Society, and a foreign member of societies abroad; he moved in the highest circles, and married a fashionable and wealthy widow. While in Bristol, he had become friends with Coleridge and Wordsworth; and later he also became the friend of Scott. But he was not much loved or respected as a man by his scientific contemporaries, who thought his ambitions were as much social as intellectual.

Dalton made his way differently; he came to Manchester, which was booming as the centre of the new cotton industry, to teach at the Dissenting Academy there, which existed to give a practically-biased education to the sons of those who for social and religious reasons could not go to Oxford or Cambridge, or to the Inns of Court in London. He joined the Literary and Philosophical Society, which was a centre of culture in the town, and soon gave up his teaching job to work full-time there, though he continued to take private pupils. This society was supported by a very different group from the fashionable crowds who flocked to the Royal Institution. They consisted of doctors and clergymen on the one hand, forming the educated elite of this new city, and of manufacturers, some of whom were extremely rich but unpolished, on the other. To this latter group science offered a way of acquiring social respectability through intellectual accomplishments; such men seem to have been interested not in the sort of science they could apply in their mills—no doubt they felt that they already knew enough of that—but in such gentlemanly and sublime sciences as natural history and astronomy. It was only a generation or two later, in the early Victorian period, that men from Manchester had the confidence to be proud of their technology in public.

Dalton did give some lectures at the Royal Institution, but he lacked Davy's verve and to Londoners he always seemed gauche and homespun; he fitted their model of a provincial Quaker. He refused to belong to the Royal Society when first nominated; though later he was elected, perhaps without his consent being sought, he was never anxious to play any role in the Society. But in Manchester he became an institution; he was justly famous for his work on colour-blindness, meteorology, and chemistry, and Mancunians flocked to his civic funeral and laboured to keep his memory green. Davy's work was just as practical as Dalton's—indeed the distinction between pure and applied science was hardly made at this time—but they moved in different social worlds, and their aspirations were different. It has even been suggested that Dalton owed some of his success in framing his atomic theory to his relative isolation from the major centres of scientific activity. Those in London or Paris were toying with more sophisticated theories of matter; but Dalton in Manchester could develop his crude but useful conception of billiard ball atoms somehow held in different arrangements, and then take little notice of metropolitan criticisms.

During Davy's Presidency of the Royal Society in the 1820s, there was for the first time a majority of active men of science —that is, those who had published scientific work—on the governing Council of the Society. But the majority of the members were still in no sense primarily men of science; though no doubt they were interested in the progress of science, and many were concerned with its application to fields such as agriculture and navigation. It was not until some twenty years later that scientific eminence became a necessary condition for admission to the Society, and when this happened there was loss as well as gain, for specialisation brings its problems and men of science no longer met politicians, clergymen and landowners on their own ground. The tensions between those who favoured an amateur and fashionable Royal Society, and those who wanted it to be a group of those who spent most of their time and energy on science, was most clearly shown in the Presidential election of 1830 when the candidates were the Duke of Sussex and John Herschel, who was, as his father had been, an eminent astronomer. The Duke was elected by a narrow majority; but this result only checked for a time the development of the Royal Society into a group of experts.

Those dissatisfied with the Royal Society because of its amateurishness and fashionableness joined with provincial critics of metropolitan sophistication and arrogance in forming the British Association for the Advancement of Science, which first met at York in 1831. This body became much more representative of the scientific community than the Royal Society. The idea of such an association had come from Germany; for in the 1820s Germany consisted of a large number of independent states, some of them very small, and therefore had no metropolis. Lorenz Oken, a speculative physiologist, had the idea of calling an annual meeting of all the German men of science, in a different city each year; foreigners also came to the meetings, and soon the idea spread abroad.

In England Vernon Harcourt, the chief organiser of the first meeting, was a clergyman and the son of the Archbishop of York; and in York interest in the sciences had been kindled by the discovery of bones in the nearby Kirkdale Cavern which seemed to provide evidence for the Flood described in the Bible. The British Association was thus not a radical body, but a representative one. It was with the founding of this Association that the word 'science' came to have its modern, restricted sense; and the word 'scientist' was coined at an early meeting, by analogy with 'artist', to describe the members. Not all those who came to the annual meetings, at different cities in Britain and ultimately even in Canada and Australia, could be properly described as scientists; there was still a continuum between the expert and the dilettante, but by the 1830s scientists had become a self-conscious group.

At the meetings of the British Association most of the work was done in various groups or 'sections', devoted to mathematics, chemistry, geology, and so on. And by the 1830s the divisions between those following the various sciences were becoming deeper. Contemporaries of Davy and Dalton sometimes seem polymaths : James Parkinson, who described Parkinson's Disease, was also the author of a standard work on fossils; Thomas Young proposed a wave theory of light, described the mechanism of the eye, tried to measure chemical affinities, and took the first steps towards deciphering the Rosetta Stone; and William Hyde Wollaston discovered a number of new metals, discussed seasickness and fairy rings before the Royal Society, and wrote a paper with the painter Sir Thomas Lawrence on why the eyes in portraits seem to follow one around. A generation later such a range became unusual; and John Herschel felt he had to choose whether or not to specialise. He chose not to, and became one of the great pundits of the mid-nineteenth century; but most of his contemporaries and their successors turned the other way, and became experts in narrower and narrower fields.

The trend towards specialisation is indicated by the rise of new journals and societies covering only one science, or by the late nineteenth century only part of a science. The journals of the original societies and of later academies modelled upon them had been general journals, covering the whole range of sciences. These publications enjoyed great prestige; but they were often very slow, the Academy of Sciences in Paris being notoriously dilatory about publishing papers. The Royal Society was quicker; but its *Philosophical Transactions* became through the eighteenth century an increasingly dignified publication, so that by 1800 it was appearing in a large quarto format and was beautifully printed on high quality paper and embellished with handsome engravings of apparatus or specimens. It was no longer the sort of publication which was likely to find its way into the hands of a provincial artisan, or even a manufacturer, tradesman or apothecary. Alongside the journals of the societies, there began to appear in the second half of the eighteenth century private journals giving more rapid publication in a cheaper format.

These journals often reprinted papers which had recently appeared in the more august pages of the journals of the Academy or the Royal Society; and also often printed full or partial translations of papers which had appeared abroad, thus fulfilling a very important task. Thus *Rozier's Journal* in France brought to the attention of Lavoisier and his contemporaries the work on gases which was going on in England; while in the opening years of the nineteenth century, *Nicholson's Journal* and the *Philosophical Magazine* kept British chemists aware of work done in France and Germany. In America, *Silliman's Journal* began publication in 1818 and was throughout the century the major vehicle for papers by American men of science; like its European equivalents, it included reprinted and translated papers. These proprietary journals were in the nineteenth century generally known by the editor's name; and if the editor was identified with some particular school or doctrine, then his journal would as a rule become filled with papers reflecting his interests or views; thus *Liebig's Annalen* published the kind of analytical work that he was getting his pupils to do at Giessen.

But the ones mentioned earlier were aimed at the whole scientific community rather than at a sect; and in their pages one finds papers by the great names of the day alongside those by men who are no longer remembered, or anyway not remembered for their science. Those on the fringe of science could hope to notice something that a more eminent man might have missed, and to publish it; and this was particularly the case in sciences where the theoretical structure was still weak, and many of the alleged facts uncertain. Thus in electro-chemistry in the opening years of the nineteenth century, the expert and the amateur were in this way on a level; and it was not until after Davy's work of 1806–7 that anybody taking up this field could not hope to make fundamental discoveries without a good deal of preliminary work. Rather later, geology became the science in which the amateur might hope to make great discoveries; and these more accessible journals enable us to reach some conclusions about which sciences were most popular at a given time.

But these journals were not specialised; although they did devote most space to the most popular sciences of the day, they covered the whole spectrum of the sciences. It was in the realm of natural history that the first society devoted to one branch of science in Britain was formed, and that the first specialised journals appeared. The society was the Linnean Society, founded in 1788 to carry on the work of the great Linnaeus whose collections had been bought in 1784 by a wealthy young naturalist, James Edward Smith. Natural history was in many ways the leading science of the late eighteenth century; travellers described and brought back specimens, and govern-ments sponsored expeditions to add to geographical and also to biological knowledge, while experts classified the materials and argued over the merits of a natural or artificial system. But the science was chiefly descriptive, and not too difficult for any educated person to take up; it had connections with medicine and with agriculture, and, through natural theology, which attempted to show the existence and wisdom of God from the study of the Creation, with religion. It was not surprising that Sir Joseph Banks should have been President of the Royal Society for the twenty years either side of 1800, for he was an explorer and natural historian and landowner. He had at his disposal private and official patronage, and could offer a career in natural history to a keen young gardener's boy.

The Linnean Society began to publish its *Transactions* in 1791; while the splendid proprietary *Botanical Magazine* had begun in 1787. This was one of the greatest periods for illustrated works on natural history, many of them both beautiful and scientifically important with pictures by such men as the Bauers, Redouté and James Sowerby. It was also the period at which geology was beginning to emancipate itself; Linnaeus had classified minerals, but now geologists were beginning to ask questions about fossils, and thus to apply an historical understanding in their science. But this is not evident in the early publications of the Geological Society, whose *Transactions* began in 1811, and which was the second major society in Britain dedicated to the pursuit of one science. Indeed it was a feature of the publications of the new specialised societies that they were as a rule empirical, bringing forward new facts rather than interpreting them. The Chemical Society actually decided that its *Journal*, which began in 1848, should not print papers of a purely theoretical character; but the other, older societies had in fact followed a rather similar course in their early years.

This was partly due to the need to make themselves respectable. Banks had favoured the Linnean and Geological societies, partly no doubt because their interests were close to his own. But in his old age he became increasingly concerned at the fragmentation of the scientific community which the setting up of specialised societies involved; and in particular he was concerned at the threat which they posed to the Royal Society, which had hitherto represented all those active in the sciences in England, or at least in London. He was particularly incensed at the proposal to set up an astronomical society; which was in the event not established until after his death.

As things turned out, the Royal Society was not threatened by specialised societies, although the forces which had brought these societies into existence also brought about the transformation of the Royal Society into a body of the most eminent members of a much larger and looser scientific community; the number of its members under thirty declined steadily, but it did not lose its vitality, and it even began to publish a less formal journal, the *Proceedings*, as well as the august *Philosophical Transactions*. But the existence of the specialised societies indicated, and contributed to, the difference between science in the nineteenth century, and what had gone before. By 1800 science was still a vocation rather than a profession in Britain, but men of science were coming to see themselves as a distinct group. In some fields, such as astronomy, where considerable mathematical knowledge was required if one was to be in any real sense an astronomer, the boundaries of the group were fairly clear; in others, such as natural history and geology, they were blurred. But during the nineteenth century, the divisions between the sciences became increasingly institutionalised. Great men might in the course of their working lives advance knowledge in various fields; and in the Royal Society and at the British Association they met those active in other fields. But the less exalted members of the scientific community increasingly had a more specialised training and performed more specialised work; and came to think of themselves as chemists or palaeontologists rather than as men of science.

The societies which we have mentioned were learned societies, dedicated to advancing knowledge; rather than professional ones, concerned with the maintenance of standards and with proper remuneration for their members. There had long been such professional societies in the medical field; and during the nineteenth century the various divisions of the engineering profession organised themselves into professional groups. Where they were working in an industry closely based upon science, their society might have some of the character of a learned society as well as a professional one. Thus the Institution of Electrical Engineers published its *Journal* from 1872, and provided an outlet for fundamental papers on electricity as well as for more technical articles.

Sometimes the distinction is hard to draw, and a society formed to pursue one of these ends may in time change its character. In his famous brief biography, *Faraday as a Discoverer*, Faraday's colleague John Tyndall described how Faraday had given up most of his 'professional' work in order to concentrate upon his researches in electricity and magnetism. By professional work, Tyndall meant consultancy and chemical analysis undertaken for a fee and, for an established man of science in the mid-nineteenth century, the fees were substantial. Thus Faraday in concentrating on research was not acting as a professional scientist as his contemporaries would have understood the term; and even today no doubt most academic scientists would regard themselves as university lecturers rather

D   97

than as scientists if asked what their profession was. Those who invite us to contemplate the process of the professionalisation of science in the nineteenth century may show us a very misty picture.

But there is no doubt that, as Faraday could have done, able men of science in the Victorian period could earn themselves a comfortable living by 'professional' work. This became increasingly clear with the passing of what we would call legislation against pollution; laws forbidding the adulteration of food, drugs, and water were made in the mid-nineteenth century, and could only be enforced if there were chemical analysts who could say definitely if the given sample was or was not adulterated. Various industries also began to make calls upon the expertise of chemists; but those who were thus consulted, or who were called upon as expert witnesses in celebrated poisoning trials, did not always inspire confidence. By the 1870s there was dissatisfaction among such professional or 'practising' chemists, because the Chemical Society did not see itself as a body issuing diplomas or certificates of competence, or in general acting as a professional body. Some unscrupulous and apparently incompetent analysts were evidently expressing interest in chemistry and getting themselves elected to the Society, and then advertising themselves as 'Fellow of the Chemical Society' to gullible industrialists; when they proved to be incapable, the chemical profession got a bad name.

The Chemical Society remained a learned society, though gradually the level of chemical knowledge expected of would-be members was raised, and nowadays the society caters for chemistry graduates. But in 1874 the public analysts decided to form their own professional society; and shortly afterwards the Institute of Chemistry was also formed to take care of the professional chemist generally. The profession of public analyst did not expand as rapidly as did other branches of chemical analysis; and straightforward, generally acceptable, techniques, giving reproducible results, had become available by about the end of the nineteenth century, so that a separate professional body for public analysts began to seem less necessary. Some of the members began to feel by the early twentieth century that the Society of Public Analysts should take more interest in the fundamentals of chemical analysis, and abandon the professional part of their function to what had become the Royal Institute of Chemistry. This was what eventually happened, but not until after the Second World War and after a good deal of impassioned debate.

So what had been founded as a professional society, breaking away from a learned society, became in its turn a learned society, filling an important role at the time when physical methods of analysis began to make rapid headway in chemistry. But meanwhile the Chemical Society and the Royal Institute of Chemistry were themselves slowly and painfully negotiating a reunion; in which an enlarged Chemical Society would unite all the academic and professional chemists, and the Royal Institute would survive as a part of this large group, taking care of professional matters.

If, therefore, we look at the community of chemists, we find that before the 1840s in Britain it would have been very hard to define. Most medical men would have had some chemical training; and anybody who had been to university would have had the chance to go to lectures on chemistry, though not many probably took advantage of this opportunity. Others went to lectures at institutions ranging from the Royal Institution to the Literary and Philosophical Societies and the Mechanics' Institutes; while others read *Nicholson's Journal* and some standard textbooks. From the 1840s there was the Chemical Society to bring them all together; and from the 1830s there had been the Chemical Section at the annual meetings of the British Association. In the 1870s this community split up; the number of journals grew too, so that even the academic chemists became inorganic, organic, or physical chemists by the end of the century. A century later, the enlarged Chemical Society represents again the community of chemists, who now mostly share a similar training but have very divergent interests. They are only a community in a rather weak sense; and naturally the scientific community is bound together with bonds a good deal weaker still.

The various parts of the community of chemists were held together to some extent in the late nineteenth century by William Crookes, with his journal *Chemical News*, which appeared every week. Its double columns of small print included reports of the meetings of scientific societies, original articles—

sometimes of great importance—and translated papers, editorials, book reviews, and correspondence. In 1869 a similar format was adopted by the new general journal *Nature*, edited by Norman Lockyer; these journals performed the function which had in the early years of the nineteenth century been performed by *Nicholson's Journal* and the *Philosophical Magazine*. The latter had indeed survived, and still does; but by the late nineteenth century, it had become a specialised journal covering parts of physics. Addresses given at the meetings of the British Association, for example, were printed in *Nature* and *Chemical News* in the week after they were delivered; while official publication in the *Report* of the British Association took much longer. A curious feature of these weekly journals is that long papers were ruthlessly chopped up into little gobbets; Victorians often had to read their science, like their novels, in parts. There can be little doubt that publications like these did keep the scientific community together, because all sorts of people read them; *Nature* also kept the boundaries of the community fluid, because it was read by numerous people who were not scientists.

The scientific community had been divided at the beginning of the century between the metropolitan and fashionable group associated with such places as the Royal Institution and the Royal Society, and those associated with the institutions of the provinces; although there were many Fellows of the Royal Society who lived outside London, at least in the summer. Even within London there was also a social division. The young Faraday, for example, when he was a bookbinder's apprentice, went with other young men of similar position to the City Philosophical Society to hear talks and join in discussions. The Royal Institution had had as one of its objectives the education of mechanics; Count Rumford, one of the founders, had believed strongly that technicians should know the scientific principles behind machines and not simply work by rule of thumb. But in the event, the Royal Institution became a centre of research supported by the wealthy who wanted to keep up in a general way with the intellectual life of the day, and who hoped for suggestions on the improvement of agriculture; and it was only through a fortunate meeting with a subscriber that Faraday was able to get a ticket to go to some of Davy's lectures, and ultimately to meet him and get a job in his laboratory.

There were journals which aimed at such people as the young Faraday and his friends; and during the nineteenth century, the number of technicians increased rapidly. The *Mechanics' Magazine* began publication in 1823, by which time Faraday was becoming an established scientist; and it ran for fifty years. In 1865 the *English Mechanic* began, and ran well into the twentieth century. These journals should not be underestimated; they sometimes contained original publications, and they covered a very wide range of sciences—astronomy for example is well represented. Their contributors were sometimes men who contributed as well to more exalted journals, for the line separating the mechanics from the men of science was not as fixed as we sometimes suppose. The mechanics' journals were chiefly concerned with the physical sciences; in the biological field there was the *Gardener's Chronicle,* which began publication in 1841, and bridged the gap between the horticulturist on the one hand, and the man of science such as Charles Darwin on the other; he used it in his studies of varieties and of natural selection.

There were therefore both social and subject divisions in the scientific community of the nineteenth century. There were also intellectual divisions; and sometimes these are revealed in the journals. These divisions are revealed not when some new journal is founded to publish papers on a new branch of science, but when there are two competing journals in the same field, dedicated to different theoretical positions. Classic cases of this situation are to be found in France and in Scotland at the end of the eighteenth century. In these countries the scientific community was more developed and self-conscious than in England, and Edinburgh and Paris were more important centres of scientific activity than London. In France, Lavoisier and his associates published their new theory of chemistry, with oxygen as its centre; papers in support of their theory appeared in *Annales de Chimie*, while adherents to the old theory of phlogiston published their papers in the *Journal de Physique.* This state of affairs survived into the second decade of the nineteenth century.

At about the same time in Edinburgh, James Hutton was

working up his theory of the earth. He believed that past changes should be explained in terms of causes acting at the present day, and shocked contemporaries when he found in geology 'no vestige of a beginning, – no prospect of an end'. Perhaps more important was his advocacy of fire, or volcanic energy, as the most important agent in geology; for this view was opposed to that of his great contemporary Werner of Freiberg, who believed that water was the primary agent. Hutton and his ally Sir James Hall published their papers in the *Transactions* of the Royal Society of Edinburgh; while supporters of Werner, of whom Professor Robert Jameson was the most prominent, formed a Wernerian Natural History Society with its own journal. Part of the reason behind the empiricism of the early specialised journals, such as those of the Geological Society and the Chemical Society, was the desire to escape divisions of this kind.

The lines drawn between the sciences have not remained constant over time, and the prestige of different sciences has also risen and fallen; so that relationships between physicists, chemists, biologists, and so on are probably a more interesting field to investigate than is the crude 'two cultures' division between arts and sciences. In the early nineteenth century, chemistry was one of the dominant sciences, and most of the domains of electricity and heat were included within it. It also succeeded in taking over much of mineralogy, which was the more readily abandoned by geologists because they were becoming absorbed in palaeontology, and thus interacting with anatomy. The coming of the principle of conservation of energy in the 1840s transformed the sciences, by bringing into prominence new relationships and creating the science of physics, into which the study of heat, light, electricity, and magnetism—which had hitherto chiefly gone with the earth sciences—readily fitted. The structures may be different in different countries; thus in France in the early nineteenth century, applied mathematics was predominant, while in England the tradition was more descriptive and empirical; so that the kind of astronomical work done in the two countries was, for example, very different. In France Laplace investigated mathematically the stability of the solar system, while in England Sir William Herschel discovered the planet Uranus and investigated the distribution of the fixed stars. Even within one country there may be different traditions; thus at the end of the nineteenth century the speculative astrophysicist Norman Lockyer may have been taken less seriously than he deserved because most contemporary astronomers in Britain were Cambridge men with a strong mathematical training, and he was not.

Very often advances seem to have been made when a man trained in one tradition works in another; that is when he migrates from one part of the scientific community to another. This may involve the crossing of a national frontier; thus in chemistry in the mid-nineteenth century, the advances in understanding of the arrangement of atoms in compounds—the so-called valency theory—seem to have been made almost entirely by those who worked for a time in a foreign country where the questions asked and the assumptions made were a little different from those they had come to accept at home. But it may involve only working at a different institution, as when Moseley left Oxford for a time to work under Rutherford at Manchester, and then returned to do his fundamental researches on atomic number just before the First World War. Again it may mean that someone trained in one science transfers his attention for one reason or another to a different science; thus in eighteenth-century Prussia, the Frenchman Maupertuis turned his attention profitably from astronomy to genetics, and in twentieth-century England R. A. Fisher made a similar pilgrimage from mathematics to genetics, and established the compatibility of the Darwinian and Mendelian theories.

With the coming of specialised journals and their proliferation during the later nineteenth and the twentieth centuries, and with the shifting of frontiers between sciences, it has become increasingly difficult for the scientist to keep up with the writings of others in his field—or with the 'literature' as it is misleadingly called. In the later part of the nineteenth century, the Royal Society undertook the task of preparing a catalogue of all the scientific papers published during the century. This vast enterprise was surprisingly successful, and the authors volumes of the *Royal Society Catalogue* were completed; but work on the subject volumes came to a stop when the First World War interrupted all such international enterprises, and were never resumed. The *Catalogue* did not list papers appearing in mechanics' publications, but its cover of journals is very wide, and it does list translations as well as original appearance.

We saw that in the eighteenth century there had been some publication of abstracts : but the great increase in the publication of these came in the late nineteenth century and in the twentieth century; without the various abstracting journals now available, nobody could hope to keep up with what is going on even in his own field. But abstracts are themselves out of date by the time they come out, and with the increasing pace of scientific publication further experiment in rapid and general communication is taking place; citation indexes and computerised searches of literature are becoming a standby for scientists, who are also circulating versions of their papers informally as 'preprints' before publication in a journal, as people did in the days of Mersenne in the seventeenth century before journals were invented.

We have thus seen the development of a scientific community of those reading the same journals and belonging to the same societies; and seen how, by the early nineteenth century, just as men of science began to become a self-conscious group, this community began to be split up into numerous smaller groups, which have themselves often later split up further. The examples we have chosen are chiefly from Britain; in France the various processes happened rather sooner, and in the USA rather later, but the pattern seems to have been not dissimilar. As philosophers and publicists have increasingly appealed to the scientific community, so its nature and boundaries have become if anything more blurred, and scientists doubtless only occasionally have any real feeling of belonging to the scientific community. We have not in this chapter dealt with one of the most important things that they do have in common, which is their training; for those who talk about normal science lay stress upon the drilling involved in scientific education, and there is no doubt that scientific degrees from different institutions have very much in common. This will be one of the things we shall look at in the next chapter, when we consider another aspect of science as a social activity; that is, science as a career.

5

SCIENCE AS A CAREER

Adam Smith in his *Wealth of Nations* of 1776 saw the division of labour leading to a separation of men of science as a group: 'In the progress of society, philosophy or speculation becomes, like every other employment, the principal or sole trade or employment of a particular class of citizens.' We have seen that to look upon science as a profession is even nowadays misleading, and for earlier times is much worse; for the scientific community does not have the power or will to impose formal rules of behaviour on its members, as doctors and lawyers do, or to control standards and remuneration. Within science there are and have long been various professions, including medicine, analytical chemistry, and engineering; but science itself is not a profession. But Adam Smith perhaps gives us a more useful clue when he writes of it as a trade or employment, though this may jar on some sensibilities. For in his day and since, an increasing number of people have made themselves a career in science; and it is to this aspect of science as a social activity that we shall now turn.

Until the nineteenth century, there was not much hope of getting rich by the pursuit of science; when the young Faraday approached Davy in 1813 to ask for employment in his laboratory, he was told that science was 'a harsh mistress, and in a pecuniary point of view but poorly rewarding to those who devoted themselves to her service'. Davy also 'smiled at my notion of the superior moral feelings of philosophic men, and said he would leave me to the experience of a few years to set me right on that matter'. Davy secured a comfortable income by marrying a wealthy widow; and by the 1830s Faraday could have devoted himself to professional work in chemical analysis and consultancy, and thereby become well off. Davy's cynicism about the moral feelings of men of science seems to have disappeared with his old age; for in his posthumously published *Consolations* he compares a career in science to one in the law or in politics, where the ambition of the eagle or the crawling power of the reptile were requisite. He added that, 'To me there never has been a higher source of honour or distinction than that connected with advances in science'; and urged men of rank to pursue this delightful and enviable road to distinction, conferring blessings and benefits upon their fellow creatures.

Although Davy was a self-made man, who had used science to attain a high social position, his presidency of the Royal Society pleased neither the gentlemen of leisure and position, nor the active members who were strongly committed to the advance of knowledge. His lectures at the Royal Institution had been enormously successful in attracting the fashionable, but they had not led to men of rank turning seriously to science. Indeed in these passages in his *Consolations,* Davy seems curiously old fashioned; for by the mid-nineteenth century the role which the gentleman-amateur or the part-time man of science could hope to play was becoming very restricted, and fundamental advances from such people were no longer to be expected; only those who had a formal training and were employed in some kind of science, could expect to get far. Amateurs could hope to provide some data for experts to work upon—Darwin relied upon information from several—but not to be at the forefront themselves.

This was not how things had been in the eighteenth century, or indeed for those who like Davy came to maturity at the beginning of the nineteenth century—though by then in France the day of the amateur was over. The great men of science of the eighteenth century were physicians, apothecaries, noblemen, soldiers, sea captains, civil servants, instrument makers, clergymen or academics; and it would be hard to draw up any typical career structure for them, as one can for most scientists of the last hundred years, who will have read for a degree in a science subject, gone on to work for a PhD, and then worked through fellowships and lectureships to a senior post in a university or research institute. While the careers of great men at any period are untypical, there were as we have already noted, far fewer ordinary men of science in the seventeenth and eighteenth century than there are today—the exponential growth has happened among the ordinary scientists.

Again in his *Consolations,* Davy sketched out what seemed to him the ideal preparation for a career in science. He advised reading for an arts degree and learning some ancient and modern languages before beginning science. There is a certain poignancy in this programme, for Davy himself had not been to university—even at the time of his death, only Oxford and Cambridge universities could confer degrees in England—but had been apprenticed to an apothecary-surgeon in the country. But the pattern he sketched was one which would have been common in the eighteenth century, at any rate for men of science destined to occupy the most prominent positions in science; it was what Sir Joseph Banks, for example, had done.

One reason for the lack of academic science in Britain down to the time of Davy's death in 1830 was the peculiar structure of the English universities; which were still, as Gibbon remarked, curiously monkish places. Fellowships were available to the best students after graduation; but these had to be surrendered on marriage. Fellows who were not resigned to a life of port and celibacy retained their office only until a college living came up, and then happily settled down to domestic bliss as a country clergyman. For Fellows were expected and usually required to take Holy Orders; just as undergraduates were required to be members of the Church of England. These rules about marriage did not apply to professorships; and from the mid-seventeenth century there were in both universities a number of scientific chairs. But professors cut little ice until the reforms of the mid-nineteenth century at Oxford and Cambridge, and received small stipends; and, in the eighteenth century, as often as not they seem to have treated their chair as a sinecure, and delivered no lectures and met no students. Power in the universities rested with the heads of colleges, who admitted the undergraduates and made the rules; they could be married, and this was a post which men of science occasionally got: thus Harvey and John Wilkins were heads of Oxford colleges in the seventeenth century.

Both Oxford and Cambridge had medical schools, though these fell into decline in the eighteenth century; and the upper reaches of the medical profession were normally open only to those who had graduated from these universities. Usually they had then gone to Edinburgh or Leyden for rather better medical training. Medical courses everywhere were very important indeed in the training of scientists; and while the most gentlemanly men of science possessed arts degrees, those who had to some extent to support themselves were very often medical men. Not all of them practised; even if one did not need to earn a living, a medical course was the best training in the non-mathematical sciences that was available. For in the eighteenth and early nineteenth centuries, the would-be doctor was made to learn a good deal of chemistry, physics, botany, zoology, and comparative anatomy as well as more direct vocational matters.

The advantage of medicine was that it was a profession, and a fashionable doctor could be sure of a large income so that it did not matter to him that science was a hard mistress. Men of science in Britain with a medical qualification included James Hutton the geologist, Hans Sloane the principle founder of the British Museum, Thomas Young who worked in optics, and W. H. Wollaston the metallurgist. All these were men of great eminence in their day; but the medical profession was socially divided, and those from the unfashionable end of it could also hope to contribute to science. It is not clear to what extent it helped a physician to contribute to science. Harvey declared that his practice had fallen off when he had made known his theory of the circulation of the blood, but he may not have been serious and in general a reputation for learning was no doubt to a man's advantage; though sometimes, as with Wollaston, it was a man's relative lack of success as a doctor that induced him to turn to science as a source of intellectual satisfaction and prestige.

In the lower ranks of the medical profession, things were rather different; and success in science might raise a man, or even his whole professional group, in status. Thus the physiological researches of John Hunter in the later eighteenth century gave him a position that surgeons had not previously enjoyed, and raised the status of surgery, as one can see from the position of surgeons in the forces who came to be regarded as officers rather than NCOs. His Swedish contemporary Scheele was no more than an apothecary's assistant when he began his work on gases and on minerals which was to make him one of the most famous men of science of the late eighteenth century; though Scheele does not seem to have raised apothecaries with him. Their contemporary Thomas Henry revolutionised the patent medicine industry when he advertised his magnesia for stomach aches as being chemically pure, rather than relying upon testimonies of the 'Mrs Smith of London writes that this medicine made her feel ten years younger' kind; he became a pillar of the Manchester Lit. and Phil., and a patron of the young Dalton.

We can learn something about making a career in science in the eighteenth century by looking at some of those who, while they cannot be considered anything like professional scientists, used their science to advance their status. Hans Sloane, who succeeded Newton as President of the Royal Society in 1727, had made his scientific reputation on his voyage to Jamaica as physician to the Duke of Albemarle in 1687–89; he described its natural history in two large folio volumes which came out in 1707 and 1725, and his collections formed the nucleus of what became the British Museum. Such voyages became a route to fame in scientific circles; the most famous voyager being Joseph Banks, who sailed on Cook's first voyage around the world, and went on to become President of the Royal Society for over forty years, and to bequeath his own collections to the British Museum. Banks needed no profession; he was a wealthy landowner. But the voyages with a scientific objective, of which he became a great patron, continued to provide opportunities for navigators or medical men to make themselves a reputation for astronomy or natural history.

Some of these sailed on official voyages like Cook's, of which we shall have more to say in the next chapter; here we need only remark that in the late eighteenth century and in the years immediately after the end of the Napoleonic wars a capacity for survey brought more rapid promotion to a naval officer; by the mid-nineteenth century this was no longer the case, and it no longer seems to have been of much advantage to a naval man that he was a good astronomer. The prime example of scientific eminence bringing rapid promotion was Henry Foster; who sailed to the Arctic with Parry, and on his return in 1827 was awarded the Copley Medal of the Royal Society, its highest honour, for his astronomical and geophysical work done there.

At that time the Duke of Clarence, later King William IV, was Lord High Admiral; and he immediately promoted Foster, and gave him command of a ship, *HMS Chanticleer*, to take on a scientific cruise. A committee of the Royal Society was speedily convened to determine objectives for the voyage, and in 1828 Foster set off to determine a series of latitudes and longitudes around the world using chronometers and making astronomical observations. A crucial part of the work was the fixing of points on the two sides of the isthmus of Panama; and here the voyage came to an unhappy end, for Foster was drowned in an accident in a canoe in 1831. His career was hardly typical, but it does illustrate how scientific ability could bring promotion at a time when the Royal Navy was being run down after the wars and a successful career could not be taken for granted.

Among medical men, Sloane's example of writing up the natural history of a region was followed by others who sailed to distant regions. Thus John Atkins, a naval surgeon who had sailed with ships accompanying a convoy on the run to West Africa to collect slaves and take them to Brazil and the West Indies, published an account of his voyage and of the regions he visited in 1735. Edward Bancroft from New England worked as a doctor in Dutch Guiana, and in 1769 published an account of its natural history. On the strength of this, and some researches on textiles done when he later went to Britain, he was elected a Fellow of the Royal Society, took a medical degree at Aberdeen, and built up a successful practice. His science had given him an entrance into influential circles, and enabled him to transform himself from a surgeon's mate into a respected physician. In the 1790s Thomas Winterbottom, who had studied medicine at Edinburgh and Glasgow and therefore occupied a higher position in the medical world, went to the new colony for freed slaves at Sierra Leone; he described the diseases endemic there on his return to England in 1796 to take over his father's medical practice at South Shields, where his experience of tropical medicine can have been of little direct use, but where he built up a prominent position for himself, and a good library which is now divided between the universities of Durham and Newcastle.

Scientific attainments and publications could thus sometimes assist a man in making his way in the world, either directly or indirectly, by leading him into association with eminent people. It seems less likely that scientific eminence was of much use in the legal or clerical professions. Eminent lawyers from Francis Bacon to Charles Lyell and W. R. Grove, a pioneer in thermodynamics, have played a part in the history of science; but it is doubtful whether any of them have had their scientific work rewarded with professional advancement.

The same seems to be true of the clergy in the eighteenth century; Stephen Hales the chemist and Gilbert White the naturalist never joined the ranks of the higher clergy, and even William Paley whose *Natural Theology* became one of the great standard texts of the next century never attained a deanery or a bishopric. In the nineteenth century things were rather different, and fame in science could lead to preferment; this seems to have been especially true for geologists, so that Conybeare and Buckland were appointed Deans, of Llandaff and Westminster respectively; while Peacock, the mathematician, became Dean of Ely. The only scientist-bishop of eminence at the end of the eighteenth century was Richard Watson, who wrote five volumes of *Chemical Essays*, published between 1781 and 1787; this became a standard work, and is particularly strong on applied chemistry. J. B. Sumner, Bishop of Chester and later Archbishop of Canterbury, published in 1816 his *Records of the Creation*, which is a respectable piece of natural theology with especial emphasis upon geology; while Bishop Stanley of Norwich became President of the Linnean Society, and FRS.

In general, it seems as though Davy was right: that down to 1830 science was as a rule a harsh mistress, in that her votaries were unlikely to become rich; but she did provide a way to social advancement for a few. To say this is not to deny that the chief attraction of science is intellectual, and that curiosity rather than careerism is the usual motive for pursuing it. But science is a career. Even in the eighteenth century, there were a few who lived by science, in contrast to those we have mentioned for whom science was an interest which might or might not forward them in their profession, or advance their social status.

Particularly in Britain, but not long afterwards in America too, the eighteenth century saw the beginning of lecture demonstrations in the sciences—a form of instruction which proved extremely popular, and which did spread a taste for science around the community. Naturally, detailed and formal courses in the sciences cannot be given in series of lectures attended with demonstrations of striking phenomena; the most that could be expected was that the audience would feel some of the excitement of discovery and of the advance of knowledge, and would come to see some of the general questions which were concerning men of science. They might remember too the more spectacular or surprising experiments demonstrated, and perhaps even what principle the experiment was supposed to elucidate.

Some of these popular lecturers were notable men of science in their own right; William Whiston the astronomer, Francis Hauksbee who worked in early electrical science, and J. T. Desaguliers whose field was mechanics and natural philosophy generally, all supported themselves by public lectures. In the early years of the nineteenth century, Davy attracted very large audiences to the Royal Institution, and thus similarly, though less directly, supported his research.

At a lower level came the itinerant lecturers, who set up shop, as it were, in different places, gave a few lectures, and moved on; among these there must have been a strong element of quackery. In between we find, for example, Thomas Wright of Durham, who was interested in astronomy, and held forth upon it to companies of the nobility and gentry. He wrote up his lectures into two elegant works, *Clavis Coelestis* and *An Original Theory*, published in 1742 and 1750; the latter is of particular interest because in it Wright made the suggestion that we see the Milky Way as a band across the heavens because the sun is part of a system of stars arranged in the form of a disc or grindstone. This was only one of his suggestions, and he later gave it up; but the idea was taken up in Germany by Kant, and later independently revived, with better evidence to support it, by the eminent astronomers William Herschel and Laplace. Wright's book was published by subscription, and so we know from the list of those who bought it that it was read by amateurs, who no doubt admired the splendid plates, rather than by astronomers.

By his lectures, books, and advice to his patrons on archaeology and landscape gardening, Wright supported him-

self and was able in later life to retire back to Durham where he
built himself a little observatory, and returned to his speculations
on the location of Heaven and Hell. It was sublime speculations
such as these, often accompanied by reflections upon the joys of
the inhabitants supposed to reside upon Venus or Jupiter, or
upon the hypothetical planets circling the fixed stars, which
made astronomy popular.

In the heavens, shown by the immortal Sir Isaac Newton and
his successors to be something like a vast and beautifully designed clock, there was excellent evidence for the wisdom of
God and the reign of universal law. In the biological realm,
the rule of law and the existence of a wise and benevolent
Designer could be made equally evident, though the demonstrations of physiological laws were messier; and doctors did
address audiences on such topics sometimes. We might note
that astronomy lectures were very rarely intended to turn people
into astronomers; even by the early eighteenth century, this
demanded both mathematical knowledge and skill in the making
and use of instruments. In fields such as electricity and chemistry,
the case was rather different; for here there were not great
general laws to contemplate at this period, but rather, curious
facts and tentative generalisations, and the promise of utility.
Here experts and laymen were much more on a level, and the
study of these sciences could be taken up without previous
training—enthusiasm was enough at the outset.

Lectures were one way of popularising science, and the good
lecturer could make a living this way; and as some knowledge
of the sciences became socially necessary, and valuable for
navigators, engineers and surveyors, so the need arose for elementary books. These might be written by men of science, sometimes the same as those who made their living by lecturing; or
they might be produced by men of letters or hacks, who could
write on whatever happened to be in demand. It was an
Italian, Algarotti, who produced an account of Newton's
philosophy 'for the use of ladies', which duly appeared in
English translation in 1739, just one year after Voltaire's
popular but more advanced account of Newtonian physics. In
England, dedicated disciples of Newton such as Benjamin
Martin and William Emerson produced numerous works popularising the physical sciences, which mostly went through a num-

ber of editions and became standard texts; these duly gave their authors a scientific reputation. In France, Fontenelle had in 1686 became famous as a result of publishing his very entertaining work on a plurality of worlds; this popularised the astronomical theory of Descartes, in the context of a discussion of the probable inhabitants of the various planets, and led to his appointment as Permanent Secretary of the Academy of Sciences. His previous reputation had been only as a man of letters. In Britain, Sir John Hill published in the 1740s and '50s a number of compilations on astronomy and natural history, and a slashing attack upon the Royal Society, who did not want a Grub Street hack among their number. Oliver Goldsmith's *Animated Nature* went through many editions, and shows how a literary man could produce a charming and useful work on natural history in the late eighteenth century.

Standard Newtonian textbooks were used in formal courses at dissenting academies where young Quakers, Unitarians, or other Nonconformists received a more practically based education than their Anglican contemporaries got at Oxford or Cambridge. Priestley and Dalton are probably the most famous men of science who taught for a time in such academies. But a glance at the library of the Roman Catholic seminary now at Ushaw, or that of the famous eighteenth-century clerical family of the Sharps, now at Durham, indicates that interest in the sciences was not limited to Protestant dissenters; and in the last years of the eighteenth century, William Wales FRS, an astronomer who had sailed with Captain Cook, taught at the Mathematical School at the Anglican foundation of Christ's Hospital in London, which included Coleridge and Charles Lamb among its pupils. What is worth remarking is that none of these people had a regular career structure, as today they would probably have. Wales' career was not typical for schoolmasters, who did not usually circumnavigate the globe. Priestley acted for some years as tutor to Lord Shelburne, and indeed did most of his best work in science then, while for another part of his life he was a full-time Unitarian clergyman; and Dalton did some school-teaching early in life, and for most of his career supported himself chiefly by taking private pupils, which seemed drudgery to his London acquaintances but which he liked.

This pattern, in which the man of science had to be an

opportunist persisted in Britain and the USA well into the nineteenth century; but there were some careers in science which
began to open during the eighteenth century. One of these was
instrument-making. In the first half of the nineteenth century,
the chemist was still expected to make his own test tubes from
lengths of glass tubing, to prepare rubber tubes from sheets of
india rubber, and to make his own litmus papers, as we see
from Faraday's fascinating book *Chemical Manipulation*. But
even he expected that the chemist would buy his balance, the
great precision instrument with which Lavoisier had so force-
fully transformed chemical theorising, from an instrument-maker.
Galileo had in 1609 made his own telescope, and soon afterwards he made a microscope too; but there were, and had been
for centuries, lens-grinders employed in making spectacles. One
of the best known to us of these in the seventeenth century was
the philosopher Spinoza.

In the second half of the seventeenth century, the telescope
and microscope became increasingly valuable in science; and
makers began to sell them to virtuosi such as Pepys. A man at
the frontier of knowledge, like Leeuwenhoek the microscopist
in the seventeenth century and William Herschel the astronomer
in the eighteenth century, might have to make his own instruments in order to see further than his contemporaries; but for
most purposes, standard instruments could be bought, or more
specialised ones made to order.

The instruments that survive from the eighteenth century are
often splendid affairs; the microscope tubes bound in tooled
leather, and the fittings all gleaming. This does not mean that
all instruments looked like this; in general scientific apparatus
is used until it is worn out, or it is taken to bits and the parts
used in new apparatus, and the instruments which survive are
untypical ones, which were probably not bought for heavy use.
The most famous instrument-maker of the eighteenth century
is probably James Watt; the story is well known of how he was
set to repair the steam engine used in lecture demonstrations at
Glasgow University, and hit upon fundamental improvements
which might be made to it. He then abandoned his career of
instrument-making for one in industry, which was untypical;
but in Birmingham he became a member of the Lunar Society,
which brought together industrialists like Boulton and

Wedgwood and men of science like Priestley and Erasmus Darwin.

Unlike Watt, most instrument-makers made instruments throughout their working lives; and such men as John Dollond and James Short made a good living from it, and acquired a scientific reputation too. Dollond made the doublet lenses of different glasses in which the coloured fringes around the image were eliminated; these fringes are inseparable from lenses made of only one kind of glass, and meant that with early optical instruments the image was always fuzzy. Newton and his successors had therefore preferred reflecting telescopes to refracting ones, because in the former there is only one lens, the eyepiece, and consequently less distortion than in a refractor with two lenses, the eyepiece and the objective. The mirror does not cause this kind of distortion, but a reflector needs to be much larger than a refractor of the same power; which is a disadvantage if the telescope has to be transported. With Dollond's lenses, refracting telescopes became, in the second half of the eighteenth century, very efficient, and James Short became the best known maker of them.

In 1762 and 1769 there were transits of Venus; it will be recalled that such occasions, when the planet passes across the face of the sun, only happen in pairs like this about once a century, and if the phenomenon is observed from distant places on the earth's surface then the distance of the sun from the earth can be computed. Telescopes which could be transported to Tahiti, St Helena, or Siberia were therefore in demand, and makers like Short did very well; he seems to have earned more than a mere academic astronomer could have done, and he also became a Fellow of the Royal Society, and published a book on the transits.

Survey instruments, of which the most famous maker in Britain was Jesse Ramsden, were also in considerable demand in the eighteenth century as maps based on triangulation came into use in Western Europe. Since the seventeenth century, when the pendulum was introduced, clocks had been a very important part of the equipment of every observatory; and clockmakers like Thomas Tompion were doing a job closely akin to scientific instrument-making. In the eighteenth century, John Harrison made the first practicable chronometer to go on voyages; he had done this in the hopes of winning, as he eventually did, the reward of £20,000 promised by Act of Parliament, but later makers of chronometers had a large market to supply when Cook and others had made the value of 'timekeepers' obvious.

In the field of chemistry, electrical machines (in which glass discs or globes whirled around to make sparks) and balances were the major items of equipment that had to be specially bought. In the early nineteenth century, W. H. Wollaston discovered a process for obtaining platinum in malleable, metallic form, rather than as a grey powder which was what his predecessors had produced. This inert metal was invaluable for chemical apparatus such as spatulas and crucibles, and Wollaston made a fortune out of manufacturing them, and was able to give up his medical practice and devote his time to research. He was hardly an instrument-maker, being a member of an eminent intellectual family; but his example shows us that it was possible to become rich making scientific equipment. His younger contemporary Charles Babbage tried to construct a kind of clockwork computer; he got back some of his expenses from the British government, but his example shows that one did not necessarily get rich by making scientific equipment.

We have seen James Watt move into engineering, and it was in the eighteenth century that engineering began to emerge as a profession. Men such as Smeaton, Rennie, the Stephensons and Telford, whom Samuel Smiles was to record with loving care, and the Brunels, who did not fit his picture of self-help so well and were not written up by him, made an important contribution to economic life in the eighteenth and early nineteenth century, with their canals, bridges, roads and railways; but in England down to the mid-nineteenth century, engineering was something to be picked up by apprenticeship rather than to be learned in formal courses. Institutions which did offer courses in engineering did not find them filled. This was in contrast with what happened on the continent of Europe; where there had been a famous mining academy at Selmecbánya in Hungary since 1735, and where in 1794 the *École Polytechnique* was set up in Paris, teaching pure and applied science—for this distinction was not at this date felt to be important.

In the English-speaking world, mechanics' institutes were launched at the end of the eighteenth century in the attempt to produce 'scientific' or 'intelligent' mechanics who would not simply work by rule of thumb but would know the principles upon which machinery worked, and would also acquire some general culture from the lectures and classes laid on. The best known of these bodies, the Royal Institution in London and the Franklin Institute in Philadelphia, soon moved beyond the ken of the mechanics and became important centres of research; in most mechanics institutes, the level of science remained low, and the lectures became increasingly devoted to general topics rather than scientific ones, while the membership was drawn from more respectable classes than the founders had expected. In short, as means of providing technical education for working men, the institutes were not on the whole very successful; and Victorian reformers looked to France or Germany for examples of how to teach science and technology. Those who, like Davy and Faraday, did rise from humble origins to an important place in science, picked up their science by something like an apprenticeship, as the humbler classes of doctors did in the eighteenth century too.

There seems to have been a difference in the expectations of those who launched mechanics' institutes in Britain and in the USA. The second and third decades of the nineteenth century were times of general and fervent belief in democracy in America, and the institutes were seen as egalitarian; knowledge was the key to self-improvement, and should be open to all who sought it, and the mechanic's job was an honourable one— provided he was a 'scientific' mechanic. In Britain, with much more specialisation of trades and social gradation, the landed and professional men who usually ran the institutes saw the mechanic's trade as useful rather than honourable, and the institutes, whether run by mechanics or *de haut en bas,* usually reinforced class distinctions. Davy, in a famous lecture at the Royal Institution, declared to his fashionable audience that the unequal division of property was the foundation of social life and economic progress. But in both countries there was the hope that by bringing technical education out into the open and giving mechanics a sound scientific training, in place of a con- servative attitude imbibed during apprenticeship, then tech- nology would become cumulative as advances soon became known and were passed on. At the Royal Institution, the

Franklin Institute, and their less well-known sister institutions, the apostles of applied science held forth.

The Royal Institution had soon become entirely a place for the wealthy and the fashionable. In the provinces, mechanics' institutes became more respectable; but there were as well 'literary and philosophical' societies, which catered for a higher stratum of society. Again, the science that was diffused was generally of a fairly low level, and the aim was gentility rather than utility; but we should remember that Davy, Faraday and Dalton were supported in their researches by such bodies, while numerous lesser men found opportunities for lecturing there, and thus supporting themselves wholly or in part by their science.

These societies were chiefly concerned with the physical sciences because of their relationship to technology. But the sublime science of physiology was not neglected, and doctors were often very prominent in literary and philosophical societies. Natural history was everywhere in the eighteenth and nineteenth centuries a very popular and edifying science, and a useful one —being associated with improvements in agriculture. Sir Joseph Banks was behind the introduction of merino sheep to improve the breed first in Britain, and then much more successfully in Australia, and behind the introduction of breadfruit trees from Tahiti to the West Indies. In natural history one finds some of the same class distinctions as in the physical sciences. There were jobs for gardeners, for plant collectors, for draughtsmen, for engravers, as well as for the experts who would classify the plants or animals; and similarly in geology there were miners and surveyors as well as mineralogists, palaeontologists, and theoretical geologists. These distinctions were not rigid, for although most gardeners or surveyors did not increase their status, it was possible to rise to the highest positions in the sciences.

Thus Hugh Miller, one of the best known of geological authors of the mid-nineteenth century, began life as a stone mason working in a quarry and becoming aware of fossils; and his older contemporary William Smith was a canal surveyor, and so gained the knowledge that enabled him to prepare the first geological map of Great Britain, which incorporated his insight that the rocks can be dated from the fossils they contain. This dating is relative rather than absolute; it gave no clue to how many million years ago the strata had been laid down, but it led to an understanding of geological history and brought time inexorably into geology. Another surveyor was A. R. Wallace who turned to natural history, and went collecting specimens first on the Amazon and then in Malaya and Indonesia. There he noticed the great differences in the fauna of the islands Bali and Lombok, the former of which has essentially Asiatic species and the latter Australian. To explain the geographical distribution of animals and plants, he hit upon the theory of evolution by natural selection, and wrote to Darwin, who had had the same idea for many years but had not yet published his views, although by then his evidence was sufficiently complete for him to be writing it up. The theory was therefore first communicated to the world in a joint paper by Darwin and Wallace, the Cambridge graduate and the surveyor who had left school at thirteen.

These were exceptional men, but the route from gardener or surveyor to distinguished, or at any rate professional, man of science was open. Men like Linnaeus, Buffon, P. S. Pallas and Sir Joseph Banks indeed collected specimens themselves, but were also very important in the patronage they possessed. On expeditions they nominated collectors, gardeners or botanists. Thus David Nelson had sailed as a gardener on Cook's last voyage, being in charge of planting useful vegetables in hopeful looking sites and of collecting specimens of unknown plants. On Bligh's voyage in the *Bounty* he was the botanist, with particular responsibility for the breadfruit trees which the ship was to collect from Tahiti and take to the West Indies. His status is indicated by his having a gardener, William Brown, under him; and in the mutiny, Nelson duly went with the Captain in the open boat, while Brown went with the mutineers. About the turn of the eighteenth century, George Caley was a gardener who wrote to Banks, and after being assigned some work at Kew Gardens was sent to New South Wales to collect specimens for Banks—his status there was semi-official. There he worked with Colonel Paterson, who was a Fellow of the Royal Society and was more interested in botany than in his military duties; and with Bligh, the Governor, who was this time being mutinied against by the officers of the New South Wales Corps. On

Caley's return to England, Banks secured for him the post of superintendent of the botanical garden at St Vincent in the West Indies where he had an unsatisfactory time; but the story shows how an ambitious gardener, the son of a farrier, could become a professional scientist—there was an informal career structure in natural history by the early nineteenth century.

Caley's career was an equivocal one, but in the next generation there were two gardener's boys who achieved great distinction in the world of British science and technology : John Gould the ornithologist, and Joseph Paxton, who built the Crystal Palace. Paxton caught the eye of the Duke of Devonshire, a keen grower of orchids and other exotic plants, for whom he built the Great Conservatory at Chatsworth which became a kind of prototype for the Crystal Palace; he built other structures, became a wealthy man, and ended his life as a Member of Parliament. Gould published some of the most splendid and enormous bird books ever produced; in 1838–40 he visited Australia, and on his return published magnificent works on the birds and mammals of that continent, his being the first full account and remaining a standard work because many of his are the first descriptions of the various species in the scientific literature. He was a 'splitter', careful to discriminate between species, and many of his distinctions are now regarded as merely separating subspecies; but his beautiful works were very serious contributions to natural history, and were, for example, used by Darwin.

Gould was no great artist, and the plates were made by Edward Lear, by H. C. Richter, and by Elizabeth Gould, his wife, until she died in 1841; they often used sketches by Gould as a basis for their plates. Mrs Gould was only one of a number of women who made a career in scientific illustration which, with translation and popularisation, was a field within science open to women in the nineteenth century. It is not until about the beginning of our century that we find, with Mme Curie, women making their mark by research in the physical sciences. The great popularisers were Mrs Marcet, whose *Conversations on Chemistry* inspired the young Faraday; and Mary Somerville, who provided advanced surveys of current work in a wide range of sciences which were valuable to men of science wanting to keep up with what their colleagues were doing. Mary

Somerville and Caroline Herschel, the sister of William Herschel who cooperated with him in his astronomical work, were elected honorary members of the Royal Astronomical Society. Later, so was Agnes Clerke whose *History of Astronomy during the Nineteenth Century* was recognised as a very valuable work of popularisation; but it was not until 1916 that ordinary Fellowship of the Society was open to women.

Many of the greatest botanical artists of the eighteenth and early nineteenth century were men, such as Redouté and the Bauers; but women were beginning to enter this field, and proved to be very good at it. There are for example some superb plates of camellias by Clara Maria Pope in Samuel Curtis' monograph on them of 1819; while the great orchid books, John Lindley's *Sertum Orchidaceum* of 1837–41 and James Bateman's *Orchidacea of Mexico and Guatemala* of 1837–43 were mostly illustrated by Mrs Withers and Miss Drake. Mrs Bury drew a splendid folio of *Hexandrian Plants* published 1831–34; and this tradition of botanical drawing combining scientific accuracy with aesthetic value is continued in our own day by for example Mary Grierson in her orchid paintings, and Stella Ross-Craig in her spiky drawings of the British flora. A career in science has been possible for women for almost as long as for men, but the fields readily open have been narrower; and, down to the twentieth century, the majority of those who have made a name for themselves in science have been European men.

In France, at the time of the Revolution, came the mobilisation of scientists, to supervise such unfortunate processes as the melting down of church bells to make cannon; and then the founding of the *École Polytechnique* in 1794. The belief there was that theoretical science would destroy superstition, and applied science would bring economic and social progress. The *École Polytechnique* was a military school, but those who taught there were civilians and were among the most eminent men of science in France—and at this date science in France was at a higher level than in any other country. Under the Napoleonic regime, the school became increasingly military in emphasis, and the teaching of pure science seems to have declined; the syllabus was very little changed for many years. Nevertheless, many of the most eminent men of science in

France in later generations went through this school, which could be said to open a career in science to its students.

Things were not entirely straightforward, however, and patronage played a role not very different from that in Britain. Laplace and Berthollet, respectively the leading astronomer and chemist in France about 1800, had houses in Arcueil near Paris, and under Napoleon enjoyed great prestige and a comfortable income from the state. They formed a little group of friends and disciples around them, and this little society published its proceedings from time to time; and they also helped each other in getting academic posts, election to the Academy of Sciences (which was at this time called the *Institut*), and so on. Just as in Britain the route to a career in science was to gain the patronage of Banks, or in Russia of Pallas, so in France to join the Arcueil group was an important step in making one's way in the world of science.

The various posts available to men of science in France did not, even in the nineteenth century, bring a large or even adequate income; and the system—to us, it seems an abuse—grew up of holding various posts in plurality, as contemporary clergymen did in England. This was known as '*le cumul*', and it meant that fewer and fewer people held more and more powerful positions; cliques, exclusiveness, a tendency to abandon science for politics, and extreme centralisation, all seem to have been features of French scientific life in the years after the battle of Waterloo and the reputation and importance of Paris as a centre of science accordingly declined, as compared to London or various German cities. But despite all this, it seems to have been true that it was in France in the revolutionary and Napoleonic periods that men of science emerged as a self-conscious and self-confident group, claiming an authority to replace that exercised by clergy or men of letters. Even if there was not yet a formal career structure for scientists, there were apparently two cultures in France at this date.

Germany and Scotland in the late eighteenth century, and on into the nineteenth, were relatively poor countries with good educational systems; and many of those trained there had to go abroad to make their living. Many went to England, which was easy for Scots and also for Germans after 1715 when the Hanoverians came to the throne; while others went to Russia, where engineers, natural historians, navigators and surveyors, and doctors could all find jobs. The medical schools of Edinburgh and Göttingen and the mining academy at Freiberg provided excellent trainings in anatomy and chemistry on the one hand, and in mineralogy on the other. The great innovation in the teaching of science came from Scotland and Germany in the 1820s, when laboratory instruction was introduced into the teaching of chemistry, and indeed made central to it.

Thomas Thomson wrote the most successful textbook of chemistry in English in the early years of the nineteenth century; it went through many editions, was even translated into French although chemistry was at this time almost a French science, and made Dalton's atomic theory generally known. Thomson then in 1813 launched a new journal, *Annals of Philosophy,* which published many important chemical papers, most notably that by Berzelius proposing the chemical symbols (H for hydrogen, O for oxygen, and so on) that we still use. At this time Thomson lived in Edinburgh; but he was then called to a chair at Glasgow, and there he made his students do practical work. He was in the medical school, and the students would have been medical students. But instead of letting them pursue their interests, he fagged them and made them do analyses for a book he was writing which would he hoped vindicate Prout's hypothesis that all chemical elements were ultimately composed of hydrogen, and that their atomic weights would therefore be integer multiples of that of hydrogen.

He or his students cooked or trimmed the results of analyses so that they fitted Prout's hypothesis very well; but Thomson was discredited when others could not reproduce his figures, and his experiment in laboratory instruction came to nothing. It was otherwise with the work of Justus Liebig, who in 1825 was appointed to a Chair at Giessen, then a small and little known German university. There he built up the most celebrated school of chemistry of the day, by concentrating on practical work, allowing students a surprisingly free hand, and publishing their work in his own journal, *Liebig's Annalen,* which became one of the most important publications in chemistry. The analysis of organic compounds had just been reduced, by Liebig among others, to a matter of relatively straightforward following of rules and manipulative techniques; so that his method of teaching was practicable. His example was soon followed at other German universities, and down to 1914 a PhD degree from Germany was almost essential for anyone wanting to follow a career in chemistry. Liebig was not the first to give any practical instruction, but he worked out how to do it for a number of students at a time—Berzelius had taken one student per annum for many years, and given them an excellent training—and thus built up a research school.

In Britain there were attempts to copy Liebig's methods, but until contact with Germany was cut off in 1914 they were relatively unsuccessful, and the PhD degree was only introduced in British universities at that time. Undergraduate courses in chemistry were offered, for example in University College, London, from its inception in the 1820s; at Oxford and Cambridge, the teaching of science suffered from the reforms of the opening years of the nineteenth century, when the work for the degrees in classics and mathematics respectively was increased, and able students wanting to get a good degree, a First in Greats or a Wranglership, had no time to spare to go to lectures on chemistry or geology. But in undergraduate teaching, even in so practical a subject as chemistry, laboratory instruction in England was usually regarded as an extra, which must be paid for if taken. The Prince Consort was largely responsible for getting A. W. Hofmann, Liebig's assistant, to come to London to take charge of the Royal College of Chemistry from 1845 to 1865. The most famous discovery made there was W. H. Perkin's preparation of the first synthetic dye; but leadership in the dye industry soon passed to Germany, where there were abundant well-trained chemists, and the Royal College under Hofmann never attracted many students, although legislation on pure food and drugs had led to a need for analysts.

It was not until about the time that Hofmann left London for Berlin in 1865 that systematic training in the sciences began to catch on in Britain. At South Kensington a complex of institutions, including the new British Museum (Natural History), the School of Mines, the Normal School for the training of teachers, and the Royal College of Chemistry were brought together on a site purchased with the profits of the Great Exhibition of 1851; the colleges eventually merged to form

Imperial College, and other scientific museums were established on the site. From the 1860s onwards the propaganda which successive presidents of the British Association had been putting out, that Britain would soon fall behind Germany and the USA if she continued to lag in scientific and technical training and to adhere to obsolete methods, began to take effect. The dramatic Prussian victories in 1870 had a salutary effect in this connection. In Manchester a great school of chemistry was begun under Roscoe; at Cambridge, the Cavendish Laboratory was founded; at Oxford, degree courses in chemistry with systematic laboratory training started; and in other new universities around the country science courses occupied a prominent place. Chemistry and zoology had by now become emancipated from medicine, and physics from mathematics, and a degree structure not unlike that of today had emerged.

In the United States, various academies and technical institutions had come into being, and American scientists were well known in the descriptive sciences, where the Coast Survey provided many jobs since it was responsible for surveying the West as well as the coast, and for precision measurements and apparatus. The diffraction gratings made by Rowland, and his determinations of various physical quantities, were celebrated. Rowland was Professor of Physics at the new Johns Hopkins University at Baltimore, founded in 1876 on the German model with emphasis placed upon research as well as teaching; the example of Johns Hopkins was soon followed by the leading universities in the USA. The famous Michelson-Morley experiment, in which they sought in vain for evidence of the earth's motion through the hypothetical æther, was performed at the Case Institute at Cleveland, Ohio in the 1880s; but this was then a place chiefly dedicated to technical teaching, and they were not sorry to see Michelson go—just as in Scotland, Aberdeen did not seem to have much regretted the departure of Clerk Maxwell. Michelson went eventually to the new University of Chicago, founded on the German model again, in 1892; and Maxwell to the Cavendish Laboratory at Cambridge. At Harvard at the turn of the century, Theodore Richards became the great authority for determining atomic weights, and was in 1901 even offered a chair in Germany; he and Michelson had both been trained in Germany, but like their contemporaries in Britain, they helped to make it unnecessary for the next generation to do likewise.

As science departments, with research and teaching, increased in number and size at institutions for higher education, so naturally the number of academic posts available and their attractiveness increased—there was both a training and a career available in science. There were also increasing numbers of posts outside universities for which a training in a science was required. The dye industry, and indeed the chemical industry generally, metallurgical industries, mining, the electrical industry, and many others all demanded trained men; and quality control became a matter of measurement and science rather than empiricism and judgement. And in government there were now many more posts for scientists. In nineteenth-century England, the Astronomer Royal had almost *been* the Scientific Civil Service, as the only government-salaried man of science; and the advice of Airy, who held the post through most of Queen Victoria's reign, was sought on an astonishing variety of questions. Right through the century there had been the ordnance survey, and from it had grown the geological survey; while there was also the hydrographic department of the Admiralty concerned with sea charts. Factory Acts required factory inspectors; then alkali inspectors were needed to check pollution from chemical plants, and public analysts to confirm the purity of food, drugs, and water supplies; while the development of standard units of length, volume, weight, and temperature required national laboratories.

By the end of the nineteenth century, the scientific community had grown enormously in size and self-assurance : men of science no longer mostly knew each other, although they would know those in their own increasingly specialised fields; and the tension between the pure scientists, working in universities or research institutions, and the applied scientists working for government, increased and led to the setting up of separate professional and learned scientific societies. But they did now have a common training, or at least a very comparable one, unlike their predecessors who had had such a wide range of backgrounds. Down to 1850 or so, there had been little formal teaching of experimental science in schools in Britain, but this gradually came in during the second half of the nineteenth century, so that by 1900 most of those coming up to university to read science would have some idea of what they were likely to be in for.

It is curious that, while Germany gave the world the idea of a university which was a centre of both research and teaching rather than a place passing on received knowledge which gets a little staler each generation, in the early years of the twentieth century she provided, in the Kaiser Wilhelm Institute, an example of the separation of research from teaching. Prominent scientists have not always been excellent teachers, especially at the undergraduate level, and have often felt that if relieved of the chore of lecturing or conducting seminars they could get on much better with their work. Davy in England was glad to be able to give up lecturing when he married his wealthy widow; but in fact he did less research of the first rank thereafter, and it could well be argued that teaching is a valuable way of clearing the mind, provided it is not the drudgery of taking students through what is already in a standard textbook. Be that as it may, the Kaiser Wilhelm Institute was intended to provide a setting in which eminent men of science could get on with their research undisturbed. It was copied abroad, most notably in Russia where such an institute was put under the control of the Academy of Sciences; this system survived the Revolution, and today in the USSR much fundamental research is done in such institutes rather than in universities. And research institutes, financed by governments, industries, or private benefactors are a feature of scientific life everywhere.

The support of scientific research required money; and in the nineteenth century governments began to assume this burden. At first it was astronomy, natural history and earth sciences which were expensive; for in the nineteenth century much physics and chemistry could be done with home-made apparatus. But in the first decade of that century, large electric batteries had to be provided by government (in France) or by wealthy patrons (in England); and the accurate measurements of physical quantities which were a feature of the Victorian period could only be done with expensive equipment. By the 1920s, physics was becoming an expensive science; and it was also becoming an affair of a team, as astronomy, the earth sciences and natural history had long been, rather than of individuals like Newton or Faraday working for the most part on their own. By the middle of the nineteenth century, the Royal Society had some parliamentary grants to give to those who needed them for their research, while expeditions, international projects to observe magnetic or astronomical phenomena, and enterprises like Babbage's computer, were financed on a case being made out for them.

Teaching as well as research is expensive in science, and if laboratory instruction is to be given then this involves both extra capital and income as compared to arts subjects. Liebig managed to extort sums—not very large ones—from the government of Hesse-Darmstadt, as we can see from his laboratory account books, which have survived. In 1889 the British government at last gave grants towards the running of the various provincial universities and university colleges; the grants were very small at first, but they kept going institutions such as the future Bristol University, which seemed on the verge of foundering despite the efforts of William Ramsay, the chemist, who was its principal, and of course they were the thin end of a wedge. Such grants meant that by the turn of the century teaching and research was being done in a range of institutions all over Britain. In the USA, similarly, the various state universities as well as the private ones were beginning to follow the example of Johns Hopkins and encouraging research as well as teaching in experimental science.

Only with government support could the opening up of careers in science become possible, except for the very few exceptions as happened in the early nineteenth century. With such support, founded on the belief that investment in science was worthwhile, there came the great growth in numbers of scientists and in rate of publication that we are now familiar with. Science became the key to a number of professions, and science courses attracted large numbers of students most of whom did not expect to pursue original research. Of those who did, the majority would be working in a team and would be applying the established methods and theories of normal science in solving relatively straightforward puzzles—just as Liebig's students had done in the 1830s. Government support thus made science an affair on a much bigger scale, and changed its character; it also had much to do with harnessing science to the economy in peace and war. The French scientists who were mobilised in the years after the Revolution to play their part in defending the Republic, were only the first of many. Our next chapter will be concerned therefore with science and government; which is the first aspect under which we shall look at science as a practical activity.

6

SCIENCE AND GOVERNMENT

The intellectual and social aspects of science have never been particularly the concern of governments, though they have on occasion persecuted or supported scientists whose views seemed to conflict with or support an ideology, and they have in many countries supported academies of sciences. But upon the whole, governments have supported science because they have seen it as a practical activity. That is, it has always been applied science which has attracted governments, and they have seen the support of science as an investment. This is not a very new idea, for governments have invested in military engineering for centuries and by the seventeenth century were beginning to support astronomy as an aid to navigation and hence to trade. To see science as a practical activity, as knowledge that will bring power, seems to have been uncommon in antiquity, and in the medieval period (although there were many technical innovations then); but is characteristic of apologists and publicists of science since about 1600.

Through most of our period the connection of science and utility was taken for granted; popular lecturers such as Davy not only gave examples of inventions and improvements which depended upon scientific knowledge, but even felt bound to point to the intellectual excitement of the sciences as something which might be forgotten by those who saw them as merely useful knowledge leading to improved techniques.

In the early seventeenth century, the line which separates natural science from natural magic had not been clearly drawn, and it was from what we would consider magic that practical results were expected; indeed this has always been the object of magic, while men of science have usually demanded an explanation as well as a recipe that works. Astrology and astronomy were thus intertwined in the early years of the seventeenth century. Copernicus had been encouraged in his astronomical work by prominent churchmen interested in the reform of the calendar; the Gregorian calendar, introduced by Pope Gregory XIII in 1582 and now in general use, was better than the Julian calendar worked out in Julius Caesar's time because it was based upon better astronomical observations. In the old calendar, the shortest and longest days and the equinoxes had got nine days away from where they should have been : by the eighteenth century, when the new calendar was introduced into Britain, the gap had grown to eleven days; and by the twentieth century to thirteen, so that the unreformed and backward Russians had their October Revolution when the rest of the world had already progressed into November, 1917.

Adoption of Copernicus' theory did not affect measurements of the length of the year, and the new calendar was theory-free. The most distinguished living astronomer by 1582 was Tycho Brahe, a Danish nobleman who had induced his King to give him the island of Hveen for an observatory. The story is that Tycho was attracted to astronomy because he found that astronomers could predict phenomena such as eclipses, and further attracted when he found they did it so badly; that is, that his interest was chiefly theoretical. But in fact his observations on stars and planets were valuable for the drawing up of accurate tables of planetary motions, and of stellar positions that would be valuable for geography and navigation. When he fell out with the King, he was invited by the Holy Roman Emperor, Rudolph II, to Prague, then the capital of the Empire, to take up the post of Imperial Mathematician.

Rudolph lived in the middle of a continent, and cannot have had much interest in navigation; his concern was with horoscopes, which he expected Tycho to cast for him. Accurate knowledge of the motions and positions of planets would lead to better predictions, and hence to better conduct of affairs of state—there were more reasons than one for a ruler to support astronomy. Tycho adopted a conservative position in astronomy, favouring the view that the earth stood still while the sun went around it, the orbits of the various planets being centred on the sun. In 1600 Kepler joined Tycho at Prague, and on Tycho's sudden death in 1601 Kepler succeeded him and had the duty of casting horoscopes and also of drawing Tycho's observations up into tables. These Rudolphine tables finally appeared in 1627; in calculating them, Kepler had used his theory that the planets all—including the earth—moved around the sun in elliptical orbits. Use of the tables did not commit one to any astronomical theory, but their value was such (because they were accurately computed from good observations) that Kepler's modified Copernican view steadily gained ground. Kepler's more directly theoretical works had been much less used and read.

Galileo was a contemporary of Kepler. Until middle age he had pursued an academic career first at Pisa and then at Padua, which was one of the leading universities of Europe. When he thought of pointing his telescope at the skies and saw the moons of Jupiter, he sagely named them the Medicean stars in flattery of the family that ruled Florence, his birthplace to which he wanted to return. He was duly appointed Court Mathematician and Philosopher; but the object seems to have been more to add glitter to a small court by having there a prominent intellectual than for any immediate purposes of utility. The use of science for prestige purposes is something which did not end with the Medici government. The same manoeuvre was made by William Herschel in the late eighteenth century, when he named his new planet (our Uranus) the Georgium Sidus after George III, and duly received a court pension; neither of the flattering names has survived, but they sewed their purpose with rulers who were anxious to play the role of the patrons of science and the arts.

Another contemporary of Kepler was Francis Bacon, who has since the middle of the seventeenth century been one of the most famous philosophers of science, particularly in the English-speaking world. He is of great interest in the context of this chapter because he was the apostle of applied science, and his is the name that has been invoked by those urging governments to give more support to science from the time of the Common-wealth to the end of the nineteenth century and even beyond. Bacon was a lawyer and not a man of science, and it seems plausible that it was his view of the common law as a higher authority than the will of kings, that made his philosophy of science and other non-legal writings popular in the mid-seventeenth century when the English Civil War broke out. His *Advancement of Learning* was a trenchant critique of current syllabuses, and his *Novum Organum* contained aphorisms—many of them memorable, like that which says that the scientist must not be a spider or an ant, but a bee—on scientific method; but it was his *New Atlantis*, describing a Utopia run by an Academy of Sciences, which caught the imagination of his successors. It led in part to the founding of the Royal Society; in Abraham Cowley's ode to the Society, Bacon is the Moses of the scientific movement who saw the promised land but did not live long enough to enter it, as Cowley and his contemporaries at the Restoration were doing.

Bacon did not only point towards a cumulative 'normal science' which could be carried on in a more democratic and collective manner than science had been hitherto; he also stressed the way in which 'experiments of light' lead to 'experiments of fruit'. For Bacon, knowledge was power; and governments like that of New Atlantis brought peace and prosperity to their citizens through their support for science. When the mechanical world view, developed in great part by Bacon's posthumous disciple Robert Boyle, prevailed during the second half of the seventeenth century, and made astrology incredible, the utility which Bacon had sketched out for science became increasingly attractive. It is worth remarking that one misjudges Bacon if one supposes that he was simply a utilitarian, or that his interests were confined to applied science; but he did believe that proper methods of enquiry systematically pursued would lead to true scientific knowledge, and that this would always be of practical use.

At the beginning of the seventeenth century men of science began to turn their attention towards magnetism, and to produce some explanation of the phenomenon which was so vital for navigators. William Gilbert of Colchester, who was President of the Royal College of Physicians in 1599, published in 1600 his experimental treatise, *de Magnete*. This put forward the crucial idea that the earth was a great magnet. Gilbert recognised that the needle did not point due north, but he believed that the variation was to the west on the coast of North America and to the east in Europe, and attributed it to the magnetic attraction of the continents. He hoped that the magnetic variation could be charted, and then used in the determination of longitude.

The finding of longitude was the great problem facing seamen in this time. They could find their latitude by taking the altitude of the Pole star, or of another star, or of the sun at noon; and there were instruments devised for this purpose. The number of navigators on long voyages who, like Columbus, relied on dead reckoning alone was declining; but to find longitude there was no instrument or observation that could be made, and the best practice was to check dead-reckoning by latitude observations. The navigator aimed to get himself into the right latitude to arrive at his destination, or to clear some formidable cape on the way, and then run down the latitude until he arrived at the port or believed himself safely past the danger. This practice added to sailing time, and therefore to the dangers of the voyage in the days of scurvy; and, in rounding Cape Horn for example, it was easy to think that one had gone far enough to the west, and therefore to turn north only to find oneself on a lee shore of Tierra del Fuego.

Even on shore the fixing of longitude was a difficult matter, but it could be done by observation of eclipses. The position of islands—even the Scilly Islands—was often dangerously wrong on charts; thus in 1707 a fleet under Sir Cloudsley Shovel was wrecked on the Scillies and nearly 2000 men were drowned, while in the Pacific it was a very difficult task to find islands described by earlier navigators. Gilbert believed that magnetic variation was constant at a given place; and that if it were charted, the seaman would be able to determine his longitude because measurements of latitude and of variation would give him two coordinates. Thus there would be only one place having a given latitude and variation.

This was a hope that survived in a modified form down to Captain Cook's day, when reliable methods of determining longitude had at last been developed; but it had been seriously shaken by Henry Gellibrand only a generation after Gilbert. He had found that the magnetic variation anywhere itself varied; we are familiar with this from maps, where the variation when the map was made is shown at the side, with a correction to apply each year. A chart of magnetic variation would soon be obsolete, and would be much more complicated than Gilbert had supposed. That the variation varied had in fact been known to the makers of compasses for some time before even Gilbert wrote his book; and they had allowed for it by sticking the needle to the bottom of the compass card askew, so that the N on the card corresponded to true north at Antwerp or wherever the compass was made, in the year in which it was made. Measurements of variation in distant places made with compasses like these could hardly be expected to be consistent if made by different seamen on different ships.

Gellibrand lectured at Gresham College in London, which had been founded with money bequeathed by Sir Thomas Gresham, the great Elizabethan merchant, and was housed in his mansion in the City of London. Courses were given there on rather more practical lines than those available in the universities in the early seventeenth century, and some surveyors and navigators seem to have profited from them. At the Restoration, those associated with Gresham College formed the nucleus of the Royal Society, which in the early days met in the college.

Many of those who, thinking along the lines of Bacon's *New Atlantis*, had drawn up plans or constitutions for a philosophical college or Solomon's House, had hoped for money from the government to run it. But Charles II had none to spare, and as we saw earlier the Royal Society had to take the form of a kind of club. Nevertheless, the improvement of astronomy was something of major national importance to any maritime power; as preachers pointed out, to know where on earth one is going it is necessary to study the heavens. In 1675 therefore Greenwich Observatory was founded, and this was an institution supported by the government. The invention of the telescope had introduced a new order of accuracy in the observation of stars and planets, making Tycho's work obsolete, as well as disclosing enormous numbers of new stars; and the perfection of pendulum clocks led to a new accuracy in the measurement of times and hence of angles. Even in Tycho's day, astronomy was a science which had needed expensive apparatus if good enough observations were to be made; as a Victorian Astronomer Royal was to point out, bad observations are worse than none at all. When astronomy ceased to be a matter of naked-eye observations, the cost of an observatory increased; and while a surprising amount of astronomy has been done since the seventeenth century in private institutions, the series of methodical observations necessary for improving navigation could hardly have been undertaken without government support.

The French government was only slightly behind the British in setting up a scientific society, and was slightly ahead in setting up an observatory; for Cassini was appointed in 1669 to work in the Paris Observatory. Because Louis XIV's government had more money than Charles II's, the French observatory was, like the Académie des Sciences, much better endowed than its equivalent in Britain. Flamsteed, the first Astronomer Royal, had to buy most of the instruments himself and never managed to get a transit instrument, fixed in the meridian, which would have given him absolute rather than relative positions of stars directly.

Greenwich had been founded to improve navigation, but at first only the groundwork of accurate observations could be done. In 1698 Edmond Halley—who in 1720 was to be Flamsteed's successor—was given command of H.M.S. *Paramour*, for a surveying voyage into the South Atlantic to determine latitudes and longitudes, and magnetic variation; he was also to look for the unknown southern continent. The voyage was less than completely successful, for there was a mutiny and the Admiralty determined that an amateur should never again be given command of a ship; but charts useful to his successors were prepared after the voyage. It was partly as a consequence of the shipwreck of Shovel's fleet that in 1714 the Board of Longitude was set up, bringing together seamen and men of science, to adjudicate proposed methods of finding longitude and to award a prize to a satisfactory one—for which they had to wait over half a century. The board was a government initiative; but there were as yet no solutions to the problem of finding longitude in prospect.

Hydrography and navigation in early eighteenth-century England were not in a particularly distinguished state; and it was not until Captain Cook's time that energetic steps were taken to rectify this state of affairs. The eighteenth century was however, a great age of expeditions having a more or less scientific character. Cassini's measurement of the arc of a degree, that is the distance subtended at the earth's surface by one degree of latitude, seemed to indicate—in accordance with Cartesian predictions—that the earth was a prolate spheroid.

That is, that its circumference around the poles was greater than that around the equator. For Cartesians, planets were immersed in vortices in a sea of æther which would squeeze them into this shape; while for Newton and his disciples, they were moving freely in a void, and their rotation would mean that they took up the form of oblate spheroids, flattened at the poles. The effect is small, but was measurable even in the early eighteenth century.

At the time of Newton's death in 1727 most Frenchmen were Cartesians, but Voltaire and Maupertuis began propaganda for Newtonian physics, which had already been accepted in France as giving the right answers, but not as a satisfactory explanation of what was happening in the world because the action at a distance involved seemed incredible. In the 1730s the Académie des Sciences sent out two expeditions to settle this question of the shape of the earth; one to Lapland under Maupertuis, and the other to the Andean region under la Condamine and Bouguer. The arc measured near the equator was longer than that measured near the pole, and this confirmed that the earth was an oblate spheroid. The expeditions also swung pendulums to determine variations in the acceleration due to gravity; such experiments were made on succeeding expeditions, for although the general question of the shape of the earth had been settled, each generation required a new accuracy in its knowledge of the exact form of the earth. This was not only for mapping, but also for direct determinations of the gravitational constant; la Condamine found that the attraction of the Andes pulled a plumb line out of true, and the pull of the mountains could therefore be compared with the pull of the earth; an estimate of the mass of the mountain could be plausibly made, and thus an estimate of the mass of the earth.

These French expeditions thus not only provided the crucial evidence for the Newtonian system—which now had to be accepted, despite the action at a distance which seemed just a feature of the world—but also marked the first attempt to find the absolute measurements of the solar system. Newton's laws allowed one for example to calculate the mass of any planet which has moons relative to that of the earth; but until the earth's mass was known in tons, the masses of the sun, the moon and all the planets could not be determined in tons. Similarly, the relative distances of planets from the sun could be computed using the laws of Kepler and Newton; but until the distance from the earth to the Sun was known in miles, the scale of the solar system was unknown. These were problems that could be solved, provided that well-equipped expeditions were sent to remote parts of the world; this demanded government assistance, and a measure of international collaboration.

Through much of the eighteenth century, there was war between various European powers, and a ship carrying a scientific party was all too likely to become the prize of a cruiser or privateer. International agreements were therefore negotiated according to which such ships were exempted from the hazards of war. These did not always answer; thus Matthew Flinders was detained for years during the Napoleonic wars at Mauritius, then a French possession, where he had landed for refreshments on his return from surveying the coasts of Australia. Similarly, reports and collections of plants and animals sent back on another ship were likely to be intercepted. Sometimes they were then lost; but they might, as some of Humboldt's specimens from South America did, fall into good hands—his were passed on to the President of the Royal Society. One problem was that scientific information was often militarily valuable; indeed Flinders seems to have been detained in part because he stood excessively upon his dignity in dealings with the French, but chiefly because he was suspected of doing a survey of Mauritius in order to facilitate an attack by the Royal Navy.

Accurate knowledge of latitudes and longitudes, natural productions, good harbours, extent of fortification, and political situation were all matters of interest to various kinds of scientists, but they were also the kind of information that any government might wish to keep from a stronger power. The line between scientific investigation and espionage was a difficult one to draw, and one can sympathise with the Portuguese authorities in Rio, for example, when they refused to allow Captain Cook or any of his team of men of science to land, on his first voyage to the South Seas.

Some expeditions involved more than international tolerance, being based on actual international collaboration. Thus la Condamine's expedition not only measured the arc of a degree on Spanish territory—Spain and France being allies at the time

—but also helped in a survey to demarcate the frontiers between Spanish and Portuguese possessions in South America. To kill two birds with one stone is a project which has always appealed to governments. From the sixteenth century, governments had backed voyages of discovery like Columbus' in the hope that they would bring to light new countries which might be colonised or be a source of profitable trade. Discovery was supposed to constitute a claim on behalf of the discoverer's sovereign to the territories found, in the absence of prior claims; and we find discoverers through the eighteenth century duly taking possession of islands and continents, leaving behind a plate or bottle recording their visit, and sailing away. The Nootka Sound Convention of 1790 between Britain and Spain made it clear that first discovery was not sufficient to constitute a claim to territory, unless it was backed by use; this point was forced upon Spain who had vague claims on the whole western coast of North America, but had now to recognise British possession of what is now called Vancouver Island.

British interest in the Pacific, or the South Sea as it was often called, went back to the Elizabethans Drake and Cavendish; in the seventeenth and early eighteenth centuries there had been buccaneers and privateers there, out for plunder but also illicitly trading with the Spanish colonists who welcomed these peaceful manifestations; and in 1740–44 came the voyage of George Anson, during which he succeeded in capturing the Manila galleon, which sailed annually between Mexico and the Philippines, loaded with bullion. This was the bright spot in an otherwise disastrous voyage, which had been planned as a raid upon Spanish commerce and colonies as part of a war and not as a scientific enterprise of any kind. Because they did not know exactly where they were, and met adverse winds and currents, the squadron had great trouble in getting around Cape Horn into the Pacific. In the ships that did get round there was terrible scurvy. Eventually there were only enough men to man one ship, the *Centurion*, out of the original six; but she carried enough guns to bring the galleon to terms after a brief battle.

Anson's voyage, the official account of which was an enormously successful book, made clear the terrible hazards of long voyages; and later discoveries only became possible when the problem of scurvy was solved, and the determination of longitude made straightforward. Both these happened in the twenty-five years after Anson's return. The scurvy had usually put an end to voyages of discovery, making it necessary to dash for a port while there were still enough men able to work the ship. The cause of the disease remained unknown, but those like Cook, who encouraged everyone to eat fresh vegetables and take such drinks as spruce beer, and who kept the ship clean, the crew busy and morale high, avoided the disease. By the end of the eighteenth century there were a range of recommended remedies, only a few of which would be thought efficacious today; notably lemon juice, which was only replaced by the less efficient lime juice on British ships in the nineteenth century.

John Byron, the grandfather of the poet, had been on Anson's voyage in the *Wager*, which had been wrecked in getting around Cape Horn; in 1764–66 he was in command of the *Dolphin,* one of the first ships to be sheathed with copper to defeat worms. Such had been the vicissitudes of Anson's voyage, that on Byron's, which was also destined for the South Sea, the crew received a bonus. In fact Byron made a circumnavigation in record time, and made no interesting discoveries except that a voyage into the Pacific was not as terrible an undertaking as had been supposed.

The Pacific was daunting because there were so few places in it where ships could find refreshments; the islands are mostly very small, and lack the wood and fresh water which were badly needed as well as fresh meat and vegetables. But on Byron's return, Wallis took command of the *Dolphin* for another voyage, and discovered Tahiti, a mountainous island well supplied with everything a crew could desire—including of course beautiful women. Wallis' purser fixed the position of Tahiti accurately by astronomical observations—there was an eclipse while they were there—so that later voyagers could find the island again.

When in 1768 Cook set out in command of the *Endeavour,* a former Whitby collier much better suited to discovery than the faster but less manoeuvrable *Dolphin*, the problem of scurvy was no longer so daunting, and the problem of finding longitude had also been solved. Cook had himself been trained in survey on the coasts of Newfoundland and Labrador, and was a good enough astronomical observer to be appointed in that capacity by the Royal Society for the voyage, along with another astronomer, Green, who died on the way home. The lunar observations made by Tobias Mayer, and worked up by the great mathematician Euler, were sufficiently accurate to allow of longitude being determined from measurements of the distance apart of the sun and moon, given suitable tables. These tables were drawn up by Maskelyne, the Astronomer Royal, and became the *Nautical Almanac*. Hadley's quadrant was also much more accurate than its predecessors, and with it sufficiently precise measurements of angles could be made so that longitude could be determined at sea.

A degree of longitude at the equator corresponds to sixty nautical miles; a new order of accuracy now became possible in determinations of longitude, for where a degree was previously as near an estimate as one was likely to get for an oceanic island, navigators now began to work to minutes of longitude at sea, and to seconds given some time ashore—though they were not always right, and there was, for example, a systematic error in all Cook's longitudes in New Zealand of between thirty and forty minutes. In his second voyage, Cook was able to correct this by further observations; and on this voyage he had also a timekeeper, or chronometer. This was a much more convenient way of determining longitude, which was readily done by comparing local time with Greenwich time shown on the chronometer. The lunar observations were still needed as a check upon the rate of going of the timekeeper; so that there were now two independent ways of determining position at sea, and safer and more direct navigation became possible.

Cook's first voyage was directed to two ends; it was one of the series, beginning with Byron's, sent into the Pacific to investigate possibilities of trade, particularly with the great southern continent, *Terra Australis Incognita*, which geographers were sure must exist to balance the continents in the northern hemisphere; and it was sent to observe the transit of Venus, so that astronomers could compute the distance of the earth from the sun. This latter aim was an international one, and numerous expeditions were sent all over the world.

In the event, as we know, the astronomical value of the expeditions was less than had been hoped; the observation of the transit was difficult to make because the black spot of Venus did not appear to make a clean ingress on to and egress from the bright face of the sun, and the computation of the results was difficult because there had been so many observers and it was impossible to know which were the best results to use in the calculations. Only in the nineteenth century, with the development of statistics, did it become possible to take all the observations into account. Eighteenth-century international expeditions had outrun the capacity of those at home to make use of their results.

This was not the case in the field of natural history. Linnaeus had worked out a system of classification which worked, so that the flora of Botany Bay could be properly described; and by the opening years of the nineteenth century his system was beginning to give way to the natural system, based upon multiple criteria. The new system met with stiff resistance, especially in Britain from the pillars of the Linnean Society; one of the innovators was S. F. Gray, whose son John Edward Gray was to be one of the most important British naturalists of the mid-nineteenth century, working at the British Museum. The work of classification went on in the Museum, at Kew Gardens, at the *Jardin des Plantes* in Paris, and at other such places, throughout the nineteenth century, and still goes on today. Such work demands large collections, for working with the natural system the taxonomist must have a large range of organisms at hand with which to compare an unknown specimen, in order to fix its relationship to known forms. While down to his death in 1820, Banks' collection was the most important in Britain, no private collections could fifty years later compare with the national ones—which included Banks', bequeathed by him to the British Museum. Government support for science in the nineteenth century necessarily included support for museums, working as far as possible on the plan of Cuvier's museum in Paris at the Revolutionary and Napoleonic period, as centres of research and teaching as well as exhibition places.

Cook's first voyage established that New Zealand was an island and not a part of the southern continent; and brought to light the eastern coast of what we call Australia. His second put an end to speculation about a habitable southern continent because he circumnavigated the globe in about latitude 60° S, getting sometimes much further south, within the Antarctic Circle, and met no continent. His third voyage, which terminated fatally for him in the Hawaian islands which he had discovered was intended to clear up another geographical problem; the existence of a North West Passage. Ever since the sixteenth century, there had been hopes for such a passage, providing a kind of polar route for ships to the Far East; attempts to get into it from the east, often through Hudson's Bay, had come to nothing, but the west coast of North America was unmapped— or rather unsurveyed, for there were maps, which often indicated a strait—and Cook tried to get in from the west.

His voyage made it extremely unlikely that there was any passage in latitudes where one would be any use; but the reports by Samuel Hearne and Alexander Mackenzie of their trips to the mouths of the Coppermine and Mackenzie rivers, where they clearly debouched into the Polar Sea, maintained interest in the passage as a geographical question if not a practical one. Cook's voyage aroused interest in the sea otters, promising a lucrative fur trade from Nootka Sound; and later voyages in search of the North West Passage opened up new areas to whalers as well as enabling cartographers to fill up the blank spaces on their maps. Cook's search for a southern continent and for a North West Passage are examples in geography of how false hypotheses which are testable can help to advance knowledge; for in this case they led to the discovery of New South Wales and of Hawaii. Cook's last voyage was specifically described as a scientific one, though no doubt commercial advantage was hoped from it in the long run. A parliamentary reward was promised, under Acts of 1745 and 1775, to those making a successful passage.

Cook's voyage on the west coast of North America was followed by Vancouver's, which was partly a survey and partly a manifestation of interest on the part of the British government in the fur trade of Nootka. As well as putting an end to any hopes of a North West Passage in temperate latitudes, because he found no entrance to one in his careful survey, Vancouver and his surgeon Menzies collected many botanical specimens, in the course of what they called their 'researches'. Cook's discovery of New South Wales was within twenty years followed by settlement at Botany Bay, or rather at the nearby Sydney Cove; Cook's companion Banks—by then President of the Royal Society—being the major promoter of this settlement. In the opening years of the nineteenth century, the coast of Australia was surveyed by Flinders, who proved it was one land-mass, rather than two great islands. He was accompanied by Robert Brown, who became one of the greatest of botanists, and the artist Ferdinand Bauer who made exquisite paintings of Australian plants which are now at last being published; while soldiers, sailors, and administrators connected with the settlement described the plants, animals, and minerals found there.

On his voyage, off what is now South Australia, Flinders met the French expedition under Baudin which was also surveying the coasts and had a number of scientists aboard; there was never a British monopoly in voyages of discovery. Baudin's relations with his experts were less happy than those of Cook or Flinders, as is clear from his journal and from that of Péron, the natural historian; but the French seem to have been more given to naming geographical features after scientists than were the British. Surveys of Australia and the Torres Strait had clearly a great importance for a power colonising the country; and at the close of the Napoleonic wars there were spare ships and men for the Royal Navy to undertake such work on a large scale. The coasts of Africa were charted as part of the programme of suppressing the slave trade, and a fair number of zoological and botanical observations were made there; the coasts of South America were also surveyed as in the wake of independence from Spain the various countries there were anxious for trade with Britain, and good charts were necessary. Darwin's voyage on the *Beagle* from 1831–36 was a part of this surveying programme in South America.

It had been arranged that Alexander von Humboldt should travel with Baudin, but when arrangements for the expedition seemed to have fallen through he went to Spain where he got the chance of visiting South America instead; like Brown, he not only observed and collected, but also generalised. No longer were natural historians interested only in new species, but in the whole aspect of the flora and fauna of a country. One of Humboldt's most suggestive essays was on the geography of plants; and Humboldt's writings inspired a whole generation of later naturalist-travellers, including Darwin. Humboldt later travelled in Central Asia; where since the eighteenth century expeditions under Russian auspices had been exploring Siberia, and also what is now Alaska. Humboldt was not only a natural historian, but was a kind of universal scientist; he had been trained as a mining engineer, and made geological observations, and he also recorded astronomical and magnetic observations. On his return to Europe he spent many years in Paris, spending his fortune in seeing his South American work superbly published, before returning to a position at the Court of Prussia. From there he worked hard for international collaboration in astronomy and the earth sciences, urging upon governments the value of simultaneous observations of magnetic and astronomical phenomena, made regularly around the world in properly equipped observatories. It is perhaps not surprising that a German should have taken the lead in this, for Germany was still itself a congeries of independent states, and science there depended on cooperation and competition between governments.

Among the most dramatic scientific expeditions were the polar voyages, in which the Royal Navy was prominent in the years after the battle of Waterloo. The prime mover behind these voyages was John Barrow, who was Second Secretary of the Admiralty from 1803 to 1845 except for a brief gap in 1806–7. After Banks' death in 1820 Barrow was the chief promoter of exploration in Britain. The purpose of the polar expeditions was still the discovery of a North West Passage; because this was recognised to be impracticable for ordinary vessels since any passage must go north of the Arctic Circle, the expeditions should perhaps count as pure rather than applied science. Part of the point of sending them was also to support territorial claims in what is now the northern part of Canada. The most successful voyage was that of Parry, who in 1819–20 wintered at Melville Island. The crew remained healthy and happy, and a large number of astronomical, magnetic, and gravitational observations were made, under Edward Sabine, a captain in the Royal Artillery who was sent out as the expedition's astronomer; for his polar work, Sabine was awarded the Copley Medal of the Royal Society, of which he was much later President. The most dramatic expedition was that of John

Franklin who went overland down the Coppermine River and on to the Polar Sea, where he and his party took to the sea and surveyed the coast, in birchbark canoes! Then came a desperate journey back, in which many of the party died; they had been inadequately supplied, partly it was believed because the Hudson's Bay Company and its rival North West Company were locked in a trade war and did not give the support they had promised. Franklin, George Back, and the expedition's surgeon, John Richardson, were saved from starvation by Indians, and survived to be lionised in London as the men who had had to eat their boots.

Later expeditions included some semi-official ones, such as John Ross' in which his nephew James Clark Ross got to the North Magnetic Pole; and George Back's, down the river which now bears his name, in search of the Rosses when they were long overdue. The younger Ross then led an expedition to the Antarctic, penetrating into the Ross Sea and getting fairly near the South Magnetic Pole, from 1839–43; his ships were essentially floating magnetic observatories, carrying observations recommended by Humboldt.

Most of the best naturalists of nineteenth-century Britain had sailed on some kind of expedition. John Edward Gray, Keeper of Zoology at the British Museum, had not, but according to his latest biographer it was a pity that he turned down the chance to go on Beechey's voyage to the north Pacific, because his work showed his lack of experience in actual field collecting. Many of the naval officers who sailed on these expeditions became as Cook had done distinguished men of science, becoming Fellows of the Royal Society and publishing or editing works on astronomy, natural history, and earth sciences. The official reports usually contained appendices of a scientific kind, and for the publication of these—usually in a sumptuous manner—government funds were forthcoming; though not to pay the experts consulted!

The seamen learnt their science by a kind of apprenticeship. Bligh, Vancouver and Flinders had sailed with Cook, Franklin with Flinders; Beechey and the younger Ross with Parry; and Belcher, a distinguished but notoriously disagreeable hydrographer, with Beechey. Richardson trained at Haslar a number of naval surgeons who became, like he was, eminent in zoology

—including T. H. Huxley. By combinations of good luck and good judgement—the luck of survey ships became proverbial—there were no great disasters on the surveying voyages of the Royal Navy down to the middle of the nineteenth century; though at the end of the eighteenth century the great French navigator la Pérouse and his ships had simply disappeared in the Pacific, where they had been wrecked upon a reef. A similar disaster befell Franklin's last expedition, which set sail in 1845 for a further attempt at the North West Passage.

When nothing had been heard of them for two years, there was some concern and rescue expeditions began in 1848; which was too late, because after in effect making the passage the ships became hopelessly iced in off King William's Island, and the crews abandoned them only to perish on the barren shores. The search for Franklin occupied much of the 1850s, and led to the coastline of northern Canada being thoroughly charted, and the existence—and impracticality—of the passage being proved.

Also, for those setting out on the search, or for voyagers generally, the *Admiralty Manual of Scientific Enquiry* was drawn up by Sir John Herschel. This is a very valuable account of the state of the descriptive sciences about 1850; its contributors included Herschel himself, Airy the Astronomer Royal, and Darwin. The book consists of a number of essays, each by an authority in his field, telling how to make observations that will be of value to science. As a guide to practice it is unique in its range; and since it went through a number of editions over more than thirty years, it can be used to follow changes in emphasis in the sciences in this period. Most of the essays are not at all technical, and are directed at ship's surgeons or at anybody interested in adding to natural history; but the section on astronomy is more forbidding, indicating that by this time knowledge of the elements of the science could be expected of navigators.

The voyages of the eighteenth century can be seen as arrangements whereby a team of technologists, the seamen, transported some scientists to a remote place where they wanted to go. By the nineteenth century, this division had begun to break down; T. H. Huxley and J. D. Hooker were ship's surgeons rather than experts from outside, but they went on to become leading biologists; Captain Foster won the Copley Medal of the Royal Society for his astronomical work, and Captain Beechey became Vice-President of the Royal Society and President of the Royal Geographical Society, as well as a Royal Academician. Darwin's Captain Fitzroy became a Fellow of the Royal Society and a prominent meteorologist—he was responsible for the first weather forecasts in Britain. So the gap between the scientists and the practical men disappeared, and these expeditions became the work of a team. Because they were concerned with questions difficult to answer rather than difficult to ask, there was a fair prospect that government support would lead to success. Such questions as: 'What is the shape of North America?', 'Does sound travel faster at very low temperatures?', and 'How does the Earth's magnetic field vary in the neighbourhood of the poles?' could be answered by expeditions.

The Victorian exploration of Africa has been likened to the space programme of our day; but the parallels are much closer between the North West Passage expeditions and the moonshots. Arctic expeditions were concerned with barren and largely useless territory, they were paid for by governments, and they were largely carried out by members of the armed forces. They came under the banner of science, but there was a large element of national prestige invested in them—one of Barrow's favourite arguments was that if the British did not open up the North West Passage, then the Russians would. There were a few technical advances which can be attributed in part to the expeditions; canned meat replaced salted meat for them, and gradually found its way into civilian use—but until the work of Pasteur and others on the processes of putrefaction, it often went bad and seamen frequently had to throw many of their tins overboard. Special instruments were designed for Arctic use, without metal knobs to which the user's fingers would have frozen; and sleeping bags were introduced on Parry's first voyage. Much was learned about nutrition, heating and ventilation in these extreme circumstances; much of Canada was explored, and the whaling industry benefited. National prestige was no doubt also achieved; it is interesting to note that when the famous British embassy under Lord Macartney went to China in 1792—with Barrow a member of the party—they took with them a letter of King George III for the Emperor, in which the King described himself as a patron of science and exploration; this was something to boast about at this time.

In the 1870s came the voyage of the *Challenger*, fitted up for deep sea research; for by this time interest had begun to shift away from hydrography, concerned with shallow seas and harbours, towards oceanography. The collections made and the observations of the team of scientists who sailed on the *Challenger* filled forty-four tomes, with another six volumes of atlas, which came out between 1880 and 1895 and represent a high point in government support for science. Oceanography was important for commerce, because knowledge of currents shortens passage times and contributes to accurate and safe navigation; but the *Challenger*'s voyage went beyond this kind of utility, and made an enormous contribution to marine zoology. With the early twentieth century, came the Antarctic expeditions of Shackleton and Scott, which had serious scientific objectives as well as being epic journeys. There is a continuous tradition from the North West Passage voyages of the early nineteenth century to the Trans-Antarctic Expedition which formed a part of the International Geophysical Year.

Naturally Britain, as a major maritime and colonial power, took the lead in voyages of exploration; and the British Museum, where able and energetic men such as Gray and Albert Günther built up the collections despite the heavy hand of officialdom, emerged as a great centre for natural history. But other countries also sent out expeditions; those from continental powers such as Russia and the United States were probably more useful than voyages into the Arctic, for both these nations were engaged in opening virgin territory which promised to be habitable and rich. The American surveys for the transcontinental railways, sumptuously published as the *Pacific Railroad Reports,* were full of natural history and geology as well as topography; there had been also American voyages of discovery, such as Wilkes' voyage in the Pacific, and Perry's to Japan, and Matthew Maury in Washington was the founder of oceanography. The mineral resources of Siberia were brought to light by Russian geologists, sometimes in cooperation with prominent foreigners such as Humboldt or Sir R. I. Murchison, and sometimes by foreigners in the Russian service. For land or sea exploration that was going to go beyond the mere recording of a route and a few unsystematic observations, government support and a team of men of science was required. This was the 'big science' of the nineteenth century, when most physics was still an affair of individuals working by themselves or with an assistant or two, and using relatively cheap equipment. Without government support, for example, Darwin would have not sailed around the world, or had available the collections of specimens he needed in forming his theory of evolution by natural selection. While not all official science was particularly full of intellectual content, and some problems were pursued for ends only loosely connected with the advance of knowledge, government support was necessary for some kinds of research, as it was for teaching.

Darwin also got information from gardeners and from stock-breeders; for the modifications which could be induced in plants and animals by selective breeding gave him a clue as to how nature might operate through natural selection. At the end of the eighteenth century, the British government had begun to take an interest in such questions, and had set up the Board of Agriculture which was to diffuse knowledge of the best practice. The leading agriculturalists of the day were Arthur Young Secretary of the Board, and his rival William Marshall, who both wrote some trenchant reports; much of the support for the Board came from improving landlords, who compiled the majority of the surveys which covered Great Britain county by county and were published in the last decade of the eighteenth century and the first decade of the nineteenth. Agricultural and horticultural progress was also encouraged from the Royal Botanical Gardens at Kew, first under the patronage of Banks, and then under the directorship of Sir William Hooker, followed by that of his son, Sir Joseph. There was also private support; in France, just before the Revolution, Lavoisier had been conducting agricultural experiments on his estates; and in England, in the early nineteenth century, at the sheep-shearing at Woburn, the Duke of Bedford played host to agricultural improvers who inspected his innovations. The agricultural research station at Rothamsted was founded in 1843 by J. B. Lawes and was supported by the profits from the sale of superphosphate fertiliser, which he had invented; it was not until Lloyd George's budget of 1909 that government money was made available for research there and for agricultural colleges.

While the British government in the eighteenth and nineteenth century was more reluctant than many others to promote science and technology directly, it not only sponsored expeditions but also promoted the Great Exhibition of 1851 and later similar exhibitions, an activity which was soon copied by other governments. The official publications connected with the Exhibition, the *Catalogue* and the *Reports of the Juries,* make fascinating bedside books today. Lyon Playfair, a scientist who entered politics, played one of the most important parts in the organisation of the Exhibition; he was one of those farsighted enough to see that Britain's lead in industrialisation could not be maintained indefinitely, and to urge that science should be supported as a key to industrial innovation. Many of the exhibits were works of art, but the majority were examples of industrial and craft products. In the various classes there were juries who chose the best exhibits, and reported on them; membership of these juries is a key to a man's standing in the world of science and technology, and the lists can be used by the historian anxious to find out who belonged to the scientific community of the mid-nineteenth century.

We find ourselves, in reading the *Reports* or the *Catalogue,* in a different world; most striking to us perhaps is that it was a world without mineral oil—locomotives were lubricated with rape-seed oil, and the animal and vegetable oil industry was of great importance. Mass production and the use of standard and interchangeable parts had also barely begun to come into use, at any rate in Europe. In most sections, the British exhibits held their own, but there were some fields where American ingenuity caught the public eye. Most notably these were the reaping machines, for reaping was still done by hand in Britain; and the Colt revolvers. The latter had played an important part in the winning of the West, for eighteenth—and early nineteenth—century firearms had been so unreliable and had taken so long to reload that they could not guarantee victory, even in a skirmish with savages—as Cook's death demonstrated. Not only could revolvers shoot six times without reloading, but they were also mass-produced; the parts were made by machinery to tolerances such that they could be assembled into a gun without further attention.

Similar processes were carried on at the arsenals making carbines for the American army; and these processes interested the British government which was at last determined to replace the musket with which the victories of Marlborough and Wellington had been won, but which was no longer adequate for the wars of the mid-nineteenth century. Other countries were already beginning to introduce rifles, which were accurate at a distance and were to make the splendid red coat and other such uniforms obsolete. With a musket, a man had an even chance of hitting an enemy on horseback at two hundred yards; there was nor question of picking off an individual, and the object was to fire volleys when one could see the whites of the enemies' eyes. The muskets were assembled in Birmingham, from parts made separately in different workshops; the parts had to be selected to fit roughly, and then filed down by the craftsmen until they fitted exactly. Similarly, after a battle a gunsmith had to work on the various damaged muskets, altering the parts of one that were to be put on to another.

In Britain, there was surplus of labour and fear of unemployment; in America there was generally a shortage of workmen, and there were not the powerful craft traditions of Europe. Labour-saving devices caught on quicker there; and with the machines for making guns, the process had become almost automatic. The tolerances were so small that any part would fit any gun; which not only led to rapid production, but meant that a damaged weapon was an immediate source of spare parts. After the Great Exhibition, Joseph Whitworth—famous for his standardisation of screw-pitches—and others were sent to the United States to investigate the making of guns there; Whitworth's report makes interesting reading, and the American example was duly followed. In fact there had been an earlier example of the use of interchangeable parts in Britain, also connected with armaments—the machinery for making pulley-blocks for ships installed by the elder Brunel at Plymouth dockyard during the Napoleonic wars; but with peace this example had not been followed. In the later nineteenth century the British government's initiative in undertaking mass production was followed in certain industries, but not nearly so widely as in the United States.

The British government was not the first to send out experts on a mission close to industrial espionage. Indeed in the early part of the nineteenth century, various foreign experts had descended upon Britain to inspect cotton mills, iron works, and railways; and their reports when published, as some have been recently, make interesting reading because technical information was not usually published. There were patent laws in most countries, but often processes were not patented because the patent was hard to defend in the courts, and competitors reading the specification might pick up valuable ideas. It was only gradually, through the nineteenth century, that governments tightened up their patent laws as a means of fostering invention. Before that a report from a knowledgeable visitor might give a clue as to what process was going on; but since a great many details would be hard to describe, the best way of getting on was to tempt workmen from another works or another country.

Liebig commented upon what he saw as the unwillingness of the British, as compared to the Germans, to support pure research; this, if true, may have gone with adherence to a vulgar Baconianism. But by the mid-nineteenth century the British government was giving support to research through grants administered by the Royal Society; and as we saw, by the 1880s government grants were going also to universities to support the teaching of science. Other countries were perhaps ahead in their official support of research, partly because of a political tradition that the government should initiate such things, which contrasts with the individualistic and liberal tradition in Victorian Britain. In America, the government was more generous in its support particularly of technology and of science requiring expensive apparatus, such as Michelson's measurement of the velocity of light.

It was really only with the twentieth century that the situation arose in Britain and America that much science was dependent upon government for its very continuance, as some natural history and earth science had been in the nineteenth century. The voyages of that epoch had provided a splendid training for some naturalists, but had perhaps distorted the development of science by diverting men away from genuinely interesting problems into fields which were low in intellectual content but where the pay was good and there was an opportunity for honours of various kinds. This was the charge made by Babbage in his *Decline of Science in England* in 1830, and made today by critics of science. Babbage wanted more money and prestige for mathematicians, whereas today's critics are often engaged in an assault upon the whole of science, which seems a less sensible enterprise.

There have always been tensions between followers of various different sciences, and readers of the doomsday literature of our day will have noticed its effect in increasing the prestige, finance and student numbers accruing to the biological sciences, at the expense of the physical ones. This tension has existed right through the history of modern science, with different sciences being in the ascendant at different times; and the possibility of distortions, careerism, and the misuse of discoveries has always been there. Science is dynamic in both its intellectual and social aspects; just as solving one problem simply raises others in theoretical science, and we need not fear that the enterprise will ever end; so the solving of one social or technological problem gives rise to others. Science will never be complete, and it will never bring the millenium. Those whose expectations of science or of governments have perhaps been unreasonable, or at least implausible in the light of history, have been excessively disappointed when science has turned out to be a very human activity, and men of science less than godlike. There is little evidence to support the view that men of science are as a class more dispassionate or benevolent than any other group—as Davy told the idealistic young Faraday at their first meeting; but there is no reason to suppose them any worse, or to condemn the whole enterprise because it did not lead to a virtuous society.

We began this chapter with the remark that governments have supported science because they have seen it as a practical activity, bringing wealth and power but at the same time generating new problems. The relations between science and technology are interesting and complicated, and it is to them that we should now turn; remembering that our distinction between pure and applied science is not one which would have occurred to many people before the late nineteenth century. The relationship is a two-way one, for improvements in apparatus have led to discoveries just as discoveries have led to new techniques. It is therefore to this aspect of science as a practical activity that we shall now turn.

7

SCIENCE AND UTILITY

The relationship between theory and practice is not a straight-forward one. Many scientists have taken it for granted, as Bacon did, that while the main reason for the pursuit of science is the urge to understand nature, the test of a scientific theory must be practical and any scientific advance will therefore lead to some advance in technology. Indeed, down to the late nineteenth century, it seems that no real distinction was generally made between what we call pure and applied science. There was much propaganda aimed at introducing science into the arts—that is what we would call crafts—in place of rule of thumb. Mechanics' institutes were set up in the early nineteenth century to turn artisans into scientific mechanics, who would know some principles of physics and thus be guided towards sensible innovations rather than will-o-the-wisps—like the perpetual motion which the young T. H. Huxley thought he had discovered and tried to explain to Faraday. Faraday duly saw him in the library of the Royal Institution, and tried to put him right. The Royal Institution was intimately concerned with applied science; its early backers seem to have been agriculturalists wanting to improve stock and yields, who got what they wanted in Humphry Davy's researches on tanning and on fertilisers. Though some were interested in manufacturing and in bettering the condition of the poor rather than in agriculture, all found what they wanted in Davy's lectures. Davy there appeared as the apostle of applied science; presenting his new discoveries, but continually drawing attention to the practical value of the sciences.

This was not a new tradition, especially in chemistry; Bishop Watson had in the late eighteenth century lectured and written on the usefulness of the science, and Glauber at the end of the seventeenth century had attributed the prosperity of Germany to the mining, metallurgical, and chemical industries. Bleaching, and therefore the whole textile industry, had in Davy's day just been transformed with the use of chlorine; a substance which had only recently been isolated by Scheele and of which the chemical status as an element was to be established by Davy. Faraday continued at the Royal Institution to pursue a practically-based course of research in the 1820s, with his studies of steels and glasses, and his analyses of by-products of gas made from whale-oil which led him to the discovery of benzene. But as his attention shifted increasingly in the 1830s into the fields of electricity and magnetism, and to the general question of the relationships between the various forces of nature, so he abandoned his directly practical researches. His electrical work laid the foundation for the electric power industry; but it was not until after his death that the first electric power stations were built. His most far-reaching contribution to applied science thus seemed for most of his lifetime to be of little usefulness. Indeed, Faraday acquired a reputation as a scientific mandarin, who had given up lucrative consultancies in applied science to concentrate his efforts upon understanding the more recondite operations of nature.

This must be one of the best examples of the value of taking a long view, and expecting the most fundamental advances in technology to come out of studies pursued in the spirit of pure curiosity. But this was not the normal pattern in technology down to Faraday's day or indeed beyond it; and the United States in the nineteenth century is the great example of a country building up a very powerful industrial economy without a corresponding base in pure science. There were competent men of science in plenty in the United States, but most were engaged in the descriptive sciences, or were involved in making accurate measurements of quantities such as atomic weights. The powerful original thought of Willard Gibbs, a pioneer of thermodynamics, found little response at home—it was only after he was recognised by leading physicists abroad that Yale even got around to giving him a salary. The Franklin Institute in Philadelphia provides a more usual pattern of interaction between science and technology than does the Royal Institution, because it did not have men such as Davy, Faraday, and Bragg associated with it.

Those who set up the Franklin Institute wanted to promote technical education, and thus invention; in particular they wanted to promote American industry, of which Philadelphia was a centre. Accordingly, they held exhibitions of manufactures, organised lectures on the sciences, and began to publish a journal. The journal began as a vehicle for popular science, directed at working men, and for a long time it continued to publish patent specifications as a guide to inventors. But soon those more interested in promoting science, and the status of scientists in America, began to exert a stronger influence within the Institute, and the journal became a vehicle for original papers on scientific and technical subjects. The Institute's government-backed investigations into the efficiency of different kinds of water wheels, and on the causes of explosions in steamships, showed the value of a methodical investigation in dealing with technical problems. These researches involved no recondite scientific principles, but were simply exercises in the application of experimental method. This is a common situation in the history of technology, whereas the application of some completely new theory or discovery is comparatively rare, and down to the twentieth century used to be even rarer.

The value of these researches was such, that by the 1860s Philadelphia industrialists were prepared to support the Institute, and its finances were at last, after nearly half a century, reasonably secure. By then there were already some significant science-based industries; notably the chemical industry, which needed a supply of trained chemists with up-to-date knowledge of the science. But even in the chemical industry much of the work that was done did not depend upon very recent discoveries or theoretical advances. Industrial chemists had been working with chemical equilibria long before Gibbs and Ostwald provided a theoretical understanding in terms of the Phase Rule and the laws of thermodynamics, just as engineers had been building high-pressure steam engines for many years before Carnot explained why the efficiency of the engine is greater if it works over a greater temperature range between the boiler and the surroundings. Often the pattern has been that a scientist has investigated a device that works, but whose principles nobody really knows. This was true of Watt and then of Carnot in their work on steam engines. Watt applied the current theory of heat, and saw how wasteful the engines in use were; he proposed a new device, the separate condenser, to save heat. In Watt's engines, some parts were always hot, while others were always cold; in older ones, much metal was being heated up and then cooled down on each stroke, and much fuel thus being wasted.

Watt's new design for engines was very difficult to carry out with the technology of the day; but when he collaborated with Boulton, whose Soho works at Birmingham were accustomed to precision work in metal, the problems proved soluble. His engine did not itself advance the science of heat, but it is an example of how a better understanding of nature led to a better machine. By the 1820s when Carnot turned to the study of engines, Watt's patent had expired and his engines had been improved upon, especially in Cornwall where there were deep tin mines to be pumped dry and no coalfields near by, and fuel economy was therefore especially important. Carnot gave an abstract analysis of a heat engine, treating it as an ideal and frictionless construction in which all changes happened slowly and reversibly. Each stroke of the engine thus became a cycle, at the end of which everything was the same as at the beginning except that some heat had been transferred from the source— the boiler—to the sink—the surroundings—and some work had been done by the engine. The terminology indicated that Carnot believed like most of his contemporaries that heat was a kind of weightless fluid. Just as more work can be got from water if the height through which it is flowing is greater, so it followed from Carnot's analysis that if the steam in the the boiler could be made hotter, by building up the pressure, the engine would be more efficient. Engineers knew this already, and the boiler explosions investigated at the Franklin Institute were a consequence of their knowledge !

The work of Carnot did not help to improve the design or safety of steam engines directly, but it ultimately gave to the study of heat a generality and elegance that had previously been lacking; it was for this reason that a wit could say that the steam engine had done more for science than science had done for the steam engine. Watt's work therefore led to an immediate improvement in engines; Carnot's was more like Faraday's, in that the application of the laws of thermodynamics in the design of machines only happened well after his death. We met Davy's work at the Royal Institution before; some was like Watt's directly useful in some practical way, while much was only useful in the long run. In the first category could come his invention of the safety lamp for coal miners, which was a splendid example of applied physics and chemistry; while in the second came his work on electrolysis, which led to the isolation of potassium and sodium, and is now a process widely used for producing useful metals such as aluminium.

The technological discovery may have repercussions upon the purer reaches of science. Thus in the course of Davy's work on the safety lamp, he demonstrated that the explosive methane and air mixture can be made to combine peacefully in the presence of platinum; and he is thus a founder of the study of catalysis, which in its turn became long after his death a valuable industrial technique, for example in oil refineries. Faraday's work on glass, in which he was trying to make a glass of very high refractive index, also much later paid off in his pure science; for it was in a sample of this heavy glass that he first detected what he called the magnetisation of light. He found, that is, that the plane of polarisation of light is rotated by a powerful magnetic field. This established another connection between the various forces of nature, and also opened the way for the theory that light was an electromagnetic radiation, and ultimately for the discovery of radio waves. Fundamental discoveries by men of genius have therefore as a rule sooner or later found a practical application. Conversely, routine practical investigations have on occasion led to advances in the theoretical understanding of nature—as when the resistance of selenium was found to vary with the light shining on it during experiments on high resistances for post office telegraphy. This was the first example of a semi-conductor to be investigated, and marks the beginning of a road leading to the transistor.

We are therefore faced, in the nineteenth century at least, with a complex situation in which one cannot say in general that science came before technology, or that techniques preceded understanding. It is for this reason that the line between pure and applied science is hard to draw. Although in particular cases it can be drawn, there appear to be no general rules and nobody can tell what will be the eventual outcome of any course of research. This is not to say that one can have no opinion as to the probable utility of some proposed investigation; but in the long term one may well turn out to be wrong.

This complexity is found much earlier, for example in the optical industry. Reading glasses with convex lenses were first made in the medieval period, and lens-grinding grew up as a craft activity. Its humble origins are shown in the very naming of *lenses*, after such a proletarian vegetable as the lentil! It was not until after some hundreds of years that men of science took an interest in how these lenses worked, and Galileo made the telescope well known and invented a kind of microscope. His contemporary Kepler, on hearing of Galileo's observations with the telescope, dropped his purely astronomical calculations for a while and turned to the study of optics. His study led to a more powerful kind of telescope than Galileo's, using two convex lenses rather than a concave eyepiece and a convex objective lens. In the hands of Snell, Descartes, and Newton optical theory in the middle years of the seventeenth century moved ahead of practice, so that telescopes were designed using optical theory rather than the theory being worked out to account for telescopes.

The investigation of arts, or craft techniques, was a prominent feature of the so-called scientific revolution of the seventeenth century. It formed part of Bacon's programme; and in Britain and France descriptions of processes were published in the late seventeenth and the eighteenth century under the auspices of scientific societies, who also discussed them at their meetings. Before Bacon's time, there had been books published on mining; the most notable being Agricola's *De Re Metallica* of 1556 and Biringuccio's *Pyrotechnia* of 1540. Miners had assembled much data for the geological sciences, but it was to be many years before much of it could be organised into a tidy form. By the mid-seventeenth century, the importance of coal as a fuel, and of metals and building stone, made geology a manifestly useful science. Anybody who could tell whether or not there was likely to be gold 'in them thar hills'—or coal, or tin, or iron—possessed valuable knowledge. Natural histories of countries, such as Plot's *Oxfordshire* of 1677, and Borlase's *Cornwall* of 1758, contain much geology and mineralogy. It was the practical importance of geology which attracted people to it; and once there, they had to wrestle with the theoretical problems of a theory of the earth, of the crystalline forms of minerals, and of the status of fossil bodies which seemed to have some analogies with the shells and bones of living creatures.

In eighteenth-century Germany and Hungary there grew up schools of mining; and Werner, one of the greatest geologists of the eighteenth century, lectured at Freiberg and attracted students from all over Europe to his course, which was particularly strong on mineralogy. This is what one would expect in a mining academy; and the next major development in geology came from a different quarter, the surveyors. The construction of canals had been a great feature of late eighteenth-century Britain, and knowledge of strata was very important for the canal builder. William Smith found that he could use the fossils embedded in the strata to characterise them; if rocks in different places contained the same fossils, then they had been laid down at the same period. Armed with this idea, he was able to prepare the first geological map of Britain; and from about 1815 the study of fossils became the dominant part of geology.

Geology was not only useful to miners and surveyors; Smith's work was first published in a work by a clergyman friend of his, intended to establish the veracity of Moses as an historian of the creation. Similarly, Werner had believed that the strata —even basalt columns—had been laid down from water, and this conclusion seemed to support belief in Noah's Flood. Conversely, more literal interpreters of Scripture were alarmed at the long time-scale demanded by the geologists, and used by freethinkers to discredit religion. The theoretical conclusions of sciences may have a usefulness well outside the science in which they were formulated; and like any other use of science—or anything else—this is a double-edged process. Certainly the palaeontologist, sharing Ezekial's vision of the dry bones coming together into living creatures at his command, became one of the most appealing of scientists in the nineteenth century; and his science was useful to biologists, for example, in giving an historical perspective in their science. In the same way, chemistry was useful in geology in providing a theory for mineralogy, which came to depend upon chemical and crystallographic analysis, and in elucidating the process of fossilisation in palaeontology. Lecturers such as Davy urged their audiences not to think only of vulgar utility in relation to science; and we should bear his exhortation in mind, although we are chiefly concerned with usefulness in the practical sense in this chapter.

The great men involved in the rise of modern science in the seventeenth century were often very practically minded, and determined to master nature by technology if the older natural magic associated with Neoplatonic philosophy failed to work. Many of them were trained in medicine, which is after all an applied science; while others were improvers, keen on increasing the yields from mines, farms, or forests. Thus Boyle, the great chemist, seems to have become interested in the sciences because he was interested in improving estates; his father was the 'great Earl of Cork', developing his estates in what was close to a colonial situation in the early seventeenth century. The Civil War gave to surveyors such as William Petty an opportunity to exercise their skills, for the lands of the disloyal were surveyed and then portioned out among the loyal on various occasions, particularly in Ireland. Boyle's interests moved towards chemistry and physics, though he also maintained a strong interest in medicine and in theology; while Petty was a founder of the science of 'political arithmetic', or statistics applied to the purposes of government.

Seventeenth-century governments knew amazingly little, by our standards, about the countries they governed. Not only did they lack the topographical and geological information that the county histories began to supply, but they knew very little about the population. In only a few parts of Europe were there full records of births and deaths; in England there were parish records, but the children of dissenters were often not christened in the parish church and there were therefore no complete lists of births. In London there were Bills of Mortality, listing numbers and causes of death in the various parishes. These had begun in the early years of the century, and were at first published during the plague epidemics, of which the outbreak in 1665 was merely the last. Long before that date, the bills had been published regularly; in each parish the cause of death was determined by a respectable widow who was paid for her duties, but was not skilled in medical science.

These London bills were first analysed by John Graunt, a

citizen and tradesman of London, in the early 1660s. He tried to compute from them the population of London, making certain reasonable assumptions about the number of women of child-bearing age, the average age at death, and so on; but in the absence of more direct evidence such a computation could be little better than a fairly well-informed guess. He was more successful in his close analysis of the Bills, where he found that certain illnesses had decreased while others had increased in the same proportion, and concluded that here the same disease was being differently described. Apart from plague deaths, he found that the causes of death were fairly steady; he noticed that deaths from venereal disease were remarkably few, and concluded that the respectable ladies could be induced to enter a more respectable cause of death on the returns. King Charles II was impressed by Graunt's work, and by the promise of the usefulness of this political arithmetic, and ordered that he be admitted a Fellow of the Royal Society. Petty was an associate of Graunt's, and carried out similar but less original researches on other vital statistics.

Political arithmetic did not prove as useful to governments as had been hoped, and it was not until the nineteenth century that governments had much statistical information available to them. One extension of Graunt's work was soon useful, and that was the development of life tables for life insurance or annuities; here the critical work was done by Halley the astronomer, using the bills for Breslau, which, unlike those in Britain, included the ages of the people who died. Halley's paper was published in 1692, that is soon after the Revolution of 1688 which had had the effect of liberalising finance and encouraging insurance. As far as governments were concerned, the development of social statistics was like any other scientific advance, neutral; on the one hand, it provided evidence upon which juster government could be based, while on the other it could help to strengthen tyranny since an ignorant government would be less efficient in repression.

Optical instrument making was not the only craft to be affected by the progress of science in the seventeenth century; clocks had like reading glasses been made for a long time before Galileo noticed the regular swinging of pendulums. It was not long before this property of pendulums was applied in horology, with the great scientists Hooke and Huygens both claiming priority for devices to ensure accurate running of clockwork. The next great improvement, the making of a chronometer small and robust enough to be carried about at sea or on land, did not depend upon any new scientific principles but was made by John Harrison, an enterprising craftsman. Harrison's work was done in the mid-eighteenth century, and by the 1760s when his final model was being tested the Industrial Revolution was well under way. In the Midlands of England, a great centre of industry, there grew up a famous group, the Lunar Society, of men interested in applied science. The most famous members were Watt, Priestley the chemist, Wedgwood the potter, and Erasmus Darwin, the medical man and scientific poet. There were other manufacturers and men of science who belonged; and the group met at the full moon, when travelling after dark was possible, to discuss science and technology.

Watt was a man clearly involved in applying science to technology, as we have already seen; Priestley's reputation in France was made at first not with his pure chemistry but with his invention of a process for making soda water; while, on the other hand, Wedgwood, whom we usually think of simply as an industrialist, was in his day regarded as a man of science too.

In a splendid example of carrying coals to Newcastle, Lord Macartney took with him some pots made by Wedgwood as part of the present to be given on the embassy to the Emperor of China. In the event, the Chinese were impressed. Wedgwood had achieved excellent and consistent results by careful attention to the composition of his ware, and to the conditions under which it was fired. The problem in the latter case was that the temperature of the kiln was clearly crucial, and yet there was no way of measuring it because at the temperature of a kiln any ordinary thermometer would melt. Control was therefore a matter of judgment and experience like other craft techniques.

After various trials, Wedgwood devised, and described to the Royal Society, a pyrometer. He found that small balls of clay contracted as they were heated; but that at a given furnace heat the contraction after a few minutes seemed complete. He assumed what seemed to be plausible on the basis of his experiments, that the contraction was linear; that the higher the temperature of the kiln or furnace, the more the ball shrank. A set of standard pieces of clay—actually they were cylinders rather than balls—could then be used to measure high temperatures. After remaining in the furnace for a few minutes, they were dropped down a groove with a scale along it; the further they dropped, the hotter the kiln had been. Some fixed points could be put on to the scale, the melting points of silver and copper for example; and it was therefore possible to calibrate the instrument in degrees Wedgwood. It was not possible to connect this scale unambiguously with the mercury-in-glass scale of ordinary thermometers, because at temperatures they measured the clay did not contract at all. Wedgwood made certain likely guesses, which turned out not very good, and thereby was able to assign values in degrees Fahrenheit to temperatures above the melting point of glass.

Until electrical methods of measuring temperature came in in the nineteenth century, there was no way of checking whether Wedgwood was right in supposing that the contraction of his clay was linear with temperature. What mattered was that the pyrometer scale should be consistent, so that if the clay cylinders fell to a certain point in the groove then the kiln was at the right temperature for the required operation; and this seemed to be the case. To ensure consistency, Wedgwood used clay from the same part of the same pit for the cylinders; and to purchasers of his pyrometer he undertook to provide the standard cylinders. Chemists and metallurgists were not completely happy with the pyrometer, because its scale could not be connected with that of ordinary thermometers and its whole basis seemed too empirical; and indeed in the end Wedgwood's assumption about the contraction of clay proved wrong—compared to the mercury-in-glass scale, the Wedgwood scale is non-linear. As a technologist has to, Wedgwood had devised a practical and cheap way of controlling a process, and had done it within a reasonably short time; these constraints did not apply to those devising a theory of heat, and a consistent method of measuring high and low temperatures so that one degree represented the same difference of temperature at the melting point of ice or of iron, was not available until almost a century after Wedgwood's experiments.

Like the later technologists at the Franklin Institute investigating water wheels and boiler explosions, Wedgwood had been using the methods of scientific investigation rather than any recent scientific theory. But this membership of the Lunar Society kept him in touch with theoretical science; he was also a Fellow of the Royal Society, but to an energetic and practical person working in the provinces this was probably less important as a forum for discussion. The Lunar Society collapsed in the 1790s as its members aged and died, and as Priestley emigrated to the more liberal United States after the sacking of his house by the last mob supporting Church and King: Priestley was a prominent supporter of the French Revolution. The younger generation did not carry on the work of the society in the same way, but Charles Darwin's ancestors were both Darwins and Wedgwoods so that he had an excellent scientific pedigree.

The Royal Society, with its more fashionable membership, was not a body indifferent to applied science; but it was agriculture and mining, associated with natural history, and navigation, associated with astronomy, rather than industry, which was of interest to the majority of the Fellows. The eighteenth century saw the introduction of new crops, such as the turnip, into farming in Britain; the breeding of better livestock, yielding more milk, meat, and wool; and the introduction of new machinery, such as the horse hoe, and the seed drill which planted seed in neat rows so that the horse hoe could weed rapidly between them. The turnip was introduced from Holland, where it had long been in use as winter fodder for cattle; and the Swedish turnip, or Swede, came from that country a little later. For summer feeding, pastures were also improved by the introduction of new kinds of grasses, and of so-called 'artificial grasses' such as clover and lucerne, which could be incorporated into crop rotations. The flooding of meadows also came into use to increase the crop of grass that would grow on them. Farmers began to grow a mixture of kinds of grasses on fields where stock was being kept, so that some was ripe at any time; and to grow on hay fields only varieties that would ripen at the same time. In 1816 George Sinclair, who worked on the Duke of Bedford's estate at Woburn, published a handsome illustrated account of the grasses that were grown there, *Hortus Gramineus Woburnensis*.

Among those who had advised Sinclair was Davy, who from his early years at the Royal Institution had given lectures on agricultural chemistry, and in 1813 had published them; this book contained an appendix by Sinclair describing experiments on grasses made at Woburn. The chief importance of the book was that Davy urged the importance of simple chemical knowledge for the farmer. He described methods of soil analysis, and recommended the use of fertilisers. Whereas the reports on counties published by the Board of Agriculture were empirical, and simply recommended practices that had worked well in places with similar soil elsewhere in the county, Davy introduced chemical theory. His lectures on agriculture had attracted large audiences, and his book too was a success, with many editions in Britain and the USA and translations into other languages. Davy had real contacts with landowners, and a genuine interest in geology going back to his childhood in Cornwall, and his book was therefore really practical rather than theoretical. It was otherwise with his successor in this field, the great chemist Liebig; who boldly entered agricultural chemistry decrying organic fertilisers and recommending soil and crop analysis, and the making up of any elements present in the desired crop and missing in the soil. Liebig's reputation made his simplicistic view widely accepted, and his fertilisers did often—especially where agriculture was generally backward—improve yields and bring some order into a complicated subject; his work represents a bold and on the whole successful incursion into a new field. We in the twentieth century, however, have learned more of the damage that can be caused in the long term when dogmatic views from the physical sciences are used as a guide in biology.

In the realm of ideas, a successful transfer from chemistry to biology was made by Gregor Mendel in his work on heredity, when he supposed that inheritance might not involve a blending of the characteristics of parents, but might be instead a matter of particulate factors—our 'genes'. Inheritance had long been studied among plants and animals; for racehorses and breeds of dogs from at least the seventeenth century in Britain for example. In the eighteenth century it became a matter of more economic importance, as the process of selective breeding was extended to sheep and cattle; this was made possible as the spread of enclosure gave the big farmers who survived more control over breeding.

Men of science puzzled themselves over such theoretical questions as whether the mother's or father's characteristics predominated in the young, and thus whether the sperm or the egg was more important in generation; improving landlords wondered whether there was a limit to the process by which the yield of milk, wool or meat from livestock could be increased. They found on the whole that there was; that too much inbreeding to develop some characteristic often led to weakness or infertility after a few generations. But the progress in stock improvement was rapid; and the same went for the cultivation of flowers, where double or curiously coloured varieties were developed by nurserymen for gardens, and exotic plants were naturalised. For Darwin and for Mendel, this practical work on cross-breeding and breeding true was of vital importance. Some of it had been done for economic reasons; and some for fun, like the flower breeding and the pigeon fancying which particularly interested Darwin, because the different varieties of pigeon looked so very different.

By the 1850s and '6os, when Darwin and Mendel published their work, there were industries which were directly using the discoveries of contemporary scientists. The synthetic dye industry came into being after Perkin's preparation of mauve in 1856, and required chemists rather than craftsmen. There is a contrast between this work and Faraday's isolation of benzene from whale oil in the 1820s; in the latter case, we have the old pattern of the scientist investigating the working or the products of a process that had been going for some time and owed little directly to science. In the synthetic dye industry, we find the problem of scaling up a chemical reaction found to go well in the laboratory so that it can be used in industry. This has become increasingly the pattern in the newer industries that have grown up since the middle of the last century. Electric power, oil refining, metallurgy and electronics have all grown up in much closer relationship to current or recent science than did the ceramic, textile, coal and iron industries of the late eighteenth centuries.

While these closer links between industry and science were being developed, the traditional relationship between science and medicine remained extremely important. Chemistry was for the most part taught in medical schools; and at the end of the eighteenth century, pharmacists in Germany began the process of upgrading themselves. Previously an apothecary had learned his trade by apprenticeship; and while occasionally, as with the great chemist Scheele, this process might produce a genuine man of science, as a rule apothecaries ranked at the bottom of the scale of 'the learned'. In the last years of the eighteenth century, Crell began journals intended to keep apothecaries up to date with chemistry; and Wiegleb and then Tromsdorff founded private pharmaceutical schools in which the apothecary received formal instruction in chemistry—especially practical chemistry—as well as learning how to carry on his trade. In 1808 Bavaria required apothecaries to have studied at an academic institution before being licensed; in 1821 Prussia counted one year at Tromsdorff's school as equivalent to two years apprenticeship, and in 1825 formally required academic study for apothecaries. At Giessen, Liebig himself taught chiefly pharmacists; though there was no great gulf between the pharmacist and the small-scale chemical manufacturer in the 1830s.

Similarly in Britain, the pressure for chemical education, and for practical chemistry especially, came through the medical schools. In Scotland, at Edinburgh and Glasgow, practical chemistry had been taught in the 1820s and the Scottish Universities Commission of 1831 recommended that such courses be compulsory for medical students. In England, the Society of Apothecaries—the main licensing body for general practitioners at the time—recommended practical chemistry from 1831, and required it from 1835. University College and King's College, London, both without much enthusiasm, taught practical chemistry, which was expensive and time-consuming compared to lecture courses; and the London hospitals acquired professors of chemistry who included some of the most eminent chemists of mid-nineteenth-century Britain. As public health began to attract increasing attention, particularly with the cholera outbreaks from the 1830s on, so the value of chemical knowledge in dealing with pollution became apparent. But most doctors did not have to prepare their own medicines, or do analyses for pathological or public health purposes, so that by the later nineteenth century chemistry was, with subjects like comparative anatomy, coming to occupy a much less important place in medical education.

Medical men in the early nineteenth century, who had learned these sciences along with their anatomy and physiology, were well equipped to undertake scientific research in a wide range of fields; but by the 1870s scientists as a rule studied one science at university and carried on with research in that discipline. Chemistry had perhaps less to give to medicine than its supporters in the early nineteenth century—when chemistry seemed the basic science—had hoped; but pharmacists and doctors undoubtedly played an important role in advancing both chemical science and chemical industry. Liebig himself invented standard apparatus through which the analysis of organic compounds became a matter of routine, and thus opened up this vast field to the well trained chemist. He then, as we saw, went on into agricultural chemistry and succeeded in getting government support in this field; for chemistry seemed to promise to solve the problems of the hungry forties and the associated unrest of 1848, and here at least in the medium term chemistry did fulfil its promises.

So far we have looked at the indirect effects of technology upon science, in providing problems, interest, and support for various sciences at various times; and in turn deriving some advantage from its connection with one or other of the sciences. There are more direct connections through apparatus, for from the days of Galileo some of the more striking advances in science have depended upon innovations in instruments, which have in turn depended upon the state of technology. Tycho Brahe in the late sixteenth century had made the most accurate measurements of the positions of the stars and planets ever achieved with the naked eye; here there were no striking innovations in instruments, which did not differ in principle from those used in the ancient world. Tycho observed with great care, and he made estimates of accuracy, correcting his readings for instrumental error; and because Kepler's tables had all been drawn up from Tycho's data alone, while older astronomical tables were a compilation from numerous observers whose accuracy differed, they were greatly superior to what had gone before. Kepler's computations, using the theory of elliptical orbits, were also better than his predecessors' who had relied upon epicycles.

With Galileo came the innovation of the telescope. He did not invent it, and he was not the first to point it at the sky, for Harriott in England had done so already. But he was the first to realise and make known the importance of this new device in astronomy. This was the first instrument to show sights never seen before; and it was soon followed by the microscope, which similarly extended the range of vision to what had previously been invisible. The telescope disclosed myriads of new stars; and in revealing the sunspots and the phases of Venus it provided evidence for the Copernican system of the world—a system which could be seen through the telescope in miniature, in Jupiter with his four moons. As everybody knows, some of Galileo's contemporaries refused to look through the instrument; no doubt in part because they believed that it was some kind of trick, producing an image of what was not there as devices with mirrors were known to do. The likelihood of optical illusion even without mirrors and lenses was well known, and it was a common dictum that the evidence of the sense of sight alone was very fallible. No other kind of evidence is possible for a star; but telescopes could have been used on terrestrial objects. Scepticism was in the event soon overcome when the Jesuit astronomers at Rome confirmed Galileo's discoveries, and Kepler provided the optical theory to explain how the telescope worked.

The image seen through early optical instruments was very poor by our standards, and this indeed provided another prop for scepticism. The image had to be creatively interpreted rather than just observed, because the lenses were made of glass that was uneven in density and hence in refractive index, and was rarely free of bubbles. The image was also blurred because of spherical and chromatic aberration; and it was only in the eighteenth century that the refracting telescope came to yield a clear image, while the compound microscope did not come into general use until the nineteenth century. But just as the telescope in its earliest days provided evidence bearing upon the Copernican theory, so the microscope in its very imperfect form could also be used to provide answers to current questions in the biological and physical sciences. Thus the circulation of the blood was demonstrated by Leeuwenhoek and Malpighi who followed it through the capillaries with microscopes; although by the time they did so, half a century after Harvey had made his theory known, there can have been few who doubted that the blood circulated.

But the microscope was not only useful in demonstrating what everybody believed already. It led also to new discoveries: of the structure of plants; the existence of spermatozoa; and the ubiquity of minute animalculae, the protozoa and bacteria. A new instrument proved to be like a new theory; it answered some questions, but posed some more. It was long before the cell theory in botany and the science of embryology were more fully developed, and the role of microorganisms, in fermentation and in disease, was understood; and these had to depend upon improved microscopes, for the questions posed by the early microscopes could only be answered with later and better ones. Through history the existence of technical frontiers, such as the limit of resolving power of microscopes, has limited scientific advance; though naturally there have also been conceptual frontiers too. The germ theory of disease, and the connection of chromosomes with heredity, did not simply arise with better microscopes; but instruments were a condition of such a theory or connection being testable, and therefore being a fully accredited part of science.

The microscope is a familiar example of such a technical frontier to historians of science; but to ordinary folk the invention of binoculars for bird-watching is perhaps more dramatic. Darwin's theory, with its stress upon the adaptation of living creatures to their environment and upon the habits and instincts of animals, encouraged a new interest in the behaviour of birds as well as in their appearance—though this had never been absent from the work of the best ornithologists. Down to the end of the nineteenth century, the naturalist aimed to collect specimens; following the maxim 'What's missed is mystery, what's hit is history', he blazed away at any bird that came within range, and tried to collect, too, a nest with fledglings in down. The death roll from such activity was often enormous; specimens had to be skinned, and the skins preserved and sent home, where the taxonomist would prefer to have several specimens so that his descriptions, and comparisons with allied species, would not be based upon the idiosyncracies of an individual bird but would represent the genuine type.

The museum, or the naturalist, could hope to exchange spare specimens, and therefore welcomed a few extra anyway; and in the field, the ornithologist could be shooting for the table as well as for the museum—early descriptions frequently tell us whether a given bird is good to eat. Only at the end of the nineteenth century did the ornithologist cease to be primarily a hunter; by then the need for conservation of rare species was apparent, and in the majority of cases there were already museum specimens of species; the problem was one of identifying a bird seen, not of describing it for the first time for science. This was a task made possible by the invention of powerful portable binoculars; which also made it possible to observe behaviour closely—dead birds do not behave interestingly. A major example of the new possibilities opened up by binoculars was Julian Huxley's paper on the courtship habits of the great crested grebe, published in 1914.

Natural history has also been held back and helped forward at various times by the technical means available for reproducing illustrations. With the coming of printing there soon followed, in the sixteenth century, the use of woodcuts to illustrate the species described. The illustrations in the herbals of this period were often superior in botanical value to the text, which was excessively deferential towards established authorities; and the woodcuts of Brunfels' and Fuchs' herbals, for example, are often reproduced and are very handsome as well as being good guides to the appearance of the plant. Animal illustrations, in the works of Gesner and of Topsell, whose *Fore-Footed Beastes* appeared in 1607, were often less successful; partly because exotic and mythical animals were depicted as well as familiar ones, which were attractively drawn. Among exotic animals, we find a kind of Gresham's Law; so that Dürer's superb but hardly realistic rhinoceros, with a little horn between its shoulder blades and splendid armour, drove out less picturesque but more accurate illustrations until well into the eighteenth century. As often seems to be the case with a new technique, the earliest woodcuts were not surpassed; those of the sixteenth century are as fine as any done later.

There is remarkable detail in the best woodcuts, but it is a process in which it is very hard to get fine details absolutely clear. For anatomical illustrations, and for clearly showing the form of fossils—where a judgement must be made on whether the fossil really is the remains of an animal or plant—woodcuts were not in general sufficiently clear. The innovation of copper plate engraving, and occasionally in scientific works, etching, was extremely important in natural history, because detailed drawings could now be circulated throughout the scientific community. Such drawings can convey more information than a paragraph of prose; and natural historians could now be surer that they were talking about the same things. A disadvantage of copper plates is that they cannot be printed on the same press as type; type is a relief process, in which what is to show black stands up from the general surface, and so are woodcuts; copper plates are engraved or etched so that what is to show black is cut deepest. The plates therefore were separately printed and as a rule were bound in separately; though occasionally one finds books in which the paper must have been put through two presses, and contains type as well as a plate.

For the seventeenth and eighteenth centuries, copper plates were the standard form of illustration used in works of science. A problem was that they were very expensive; the cost of a detailed plate of large quarto size for the *Philosophical Transactions* in the 1810s was about twelve pounds, which was a large sum of money in those days. A further problem was that the engraving of the plates put a craftsman between the man who drew the specimen and the printed picture. Audubon's famous pictures of American birds lost some of their freshness as compared to his original watercolours even though they were printed in Britain by the Havells, who were one of the leading scientific engravers of the day. To overcome these difficulties, some scientists simply had to teach themselves engraving; and among those who did this very successfully in the eighteenth century was Mark Catesby, author of the standard *Natural History of Carolina*, who even himself coloured the early plates, because he could not afford colourists. The great illustrated works on natural history right through the nineteenth century were hand-coloured, and it is only in our day that colour printing has reached a standard comparable to that of hand colouring but less variable, for no two hand-coloured illustrations are exactly alike.

In the opening years of the nineteenth century there came a new breakthrough in illustration, with the perfection of two processes—here again the early work done with them has not been surpassed. The first of these was lithography. The picture is drawn on a stone with a greasy crayon; the stone is wetted, and then inked when the oil-based ink sticks to the greasy lines but not to the wet stone; the picture can then be printed from the stone. The engraver is thus eliminated; the artist could now do his own pictures directly on stone, and get them printed. The process was much cheaper than copper plates; when Buckland published his *Reliquiæ Diluvianæ*, describing the remains of animals found in caves, in 1823, he used copper plates prepared by the Royal Society for a shorter article for most of the illustrations, but for those required new for the book he used lithography. A nice example of a lithographed work was J. S. Miller's *Crinoidea* of 1821, describing fossil stalked star-fishes; some of his plates are 'exploded', for just as Alexander von Humboldt was demonstrating to geographers that maps could display more than just topography, so zoological illustrators were using illustrations to do more than show what an animal looked like. Some, using devices like exploded drawings, were moving towards diagrams; while others achieved greater naturalism by showing the animal in its habitat in a characteristic position, rather than stiffly posed on a mossy stump in a studio.

The other technique was wood engraving, done on the end grain of boxwood; the greatest exponent of this was Thomas Bewick, whose pictures particularly of birds are well known and very charming. William Yarrell's *British Fishes* and *British Birds*, of 1836 and 1843, were authoritative works illustrated with wood engravings, which were much cheaper than copper plates and yet included fine detail. Wood engravings could be set on the same page as type, making for more convenient reference. Soon after the middle of the nineteenth century came the use of photography for scientific illustration; in many cases this could replace an engraving, but one cannot photograph a plant with buds, flowers and fruit on it simultaneously, or show a typical specimen rather than an individual. So whereas copper plates replaced woodcuts for scientific illustration in the seventeenth and eighteenth centuries, there have been in the nineteenth and twentieth centuries a number of techniques available, and in each publication the choice must be made on the grounds of cost and suitability.

Copper plates were also used to illustrate apparatus; and just as innovations in illustration were important in natural history, so innovations in apparatus were critical in the history of chemistry. Distillation, first practiced in medieval Sicily, changed not only drinking habits but also theories of chemistry; for heating a substance and collecting what came off became the standard method of chemical analysis. Lavoisier's chemical textbook, of which the English version came out in 1790, had many handsome plates of apparatus; for Lavoisier hoped that if anybody followed his methods, he would also come to his conclusions about the nature of combustion, and believe in oxygen instead of phlogiston.

Chemists had a difficult time in Lavoisier's day, because their glass was neither mechanically strong nor chemically inert; a great problem was that if glassware was to stand up to being heated, it had to be thin; whereas if it was not to be impossibly fragile, it had to be thick. Glassware connected using tubes and corks cracked as the corks swelled, and pieces of apparatus were therefore 'luted' together with cements of various composition. The chemical reactivity of the glass was often a nuisance, as for example in early work on electrolysis; for when an electric current was passed through water other products besides oxygen and hydrogen were found. It was not until Davy in 1806 used water distilled in silver apparatus and electrolysed *in vacuo* in vessels of agate and gold that he obtained no byproducts, and could therefore clearly state his electrochemical theory. Vessels of platinum became a standby for chemists during most of the nineteenth century; and fortunately for them, by the time of the metal became fashionable for jewellery, and therefore ruinously expensive, there were other inert substances available for making crucibles and spatulas.

In the 1820s Faraday wrote his only full-length book, *Chemical Manipulation*; which is a guide to the best practice, telling the reader how to carry out chemical processes in general rather than any particular reaction. From it we can see how to weigh, how to distill, how to recrystallise, and so on; and we see that just at this time rubber tubing, called by Faraday 'Caoutchouc connectors', was coming into use to join different pieces of apparatus and thus provide the flexibility that Lavoisier's luted joints had not had. In the 1820s one had still to make the tubes from sheets of india rubber; just as one had to make test tubes from lengths of glass tubing. Faraday did not describe a condenser for use in distillation experiments; the kind of condenser that we now use is named after Liebig, and was one of the innovations made by him in the 1830s that enabled him to turn the analysis of organic compounds into a matter of routine. Later still the apparatus we use for volumetric work came into use; the burette—originally a kind of jug—was designed in a practical and accurate form by Mohr, and titrations which had previously been rather rough and ready were by the latter part of the nineteenth century an accepted procedure.

Chemists throughout the nineteenth century had to be good with their hands, and ready to make much of their apparatus themselves, or to turn things to new uses. But close on the heels of the chemist came, as we have seen, the instrument-maker, and increasingly as the century went on pieces or sets of apparatus could be bought ready made. For balances and electrical machines this was true before 1800; and the range steadily increased to include glassware, litmus papers, filter papers and bunsen burners. It was not until 1860 that chemistry had to borrow from physics a new analytical tool, the spectroscope; chemists had previously used electric batteries, but electricity was generally accounted a branch of chemistry, and spectroscopy was the first of the physical methods of analysis that have since steadily made chemistry more accurate and less messy. Here as elsewhere, there was both a technical and an intellectual barrier to cross; the new method had to be as reliable as accepted ones, and chemists had to overcome their suspicions that physicists were engaged in a plot to take over their science, leading it back to the days when men of science were satisfied with explanations in principle only, and were scornful of those who got their hands dirty. Innovation of this kind in apparatus is not very different from innovation in industry, and provokes similar reactions.

Major innovation in apparatus, as in other parts of science, is again a double-edged process. A new theory, such as Darwin's or Newton's, sent men of science in every field looking for evolutionary histories, or for equations in terms of central forces. Similarly, a new piece of apparatus becomes a new toy which everybody wants to play with; sometimes making new discoveries, and sometimes diverting themselves from a field in which they were more profitably engaged—as we can see with benefit of hindsight; to distinguish these two possibilities at the right time is a very difficult matter. Thus enormous electric batteries were built at the Royal Institution in London, and at the Institut in Paris, following Davy's isolation of potassium in 1807; this was part of an international struggle, for the Napoleonic wars were going on at the time. But in the event neither battery led to great discoveries, and the alkaline earth metals were isolated neither by Davy nor by Gay-Lussac, but by Berzelius in Sweden who had no very expensive apparatus but who hit upon the idea of using mercury rather than platinum for his cathode. One can find examples both of new kinds of apparatus making possible new discoveries, and of it acting as a straitjacket because complex apparatus is inflexible. Simple equipment—sealing wax, string and glass tubes are the proverbial paradigm—can be turned to various uses, while a new device, which the experimenter does not perhaps fully understand, tends to be used only for the function for which it was designed. There is no simple answer; but we should not be entirely unsympathetic to those astronomers who did not want to waste their time looking through Galileo's telescope.

We have now followed the three strands of the history of science, viewing it as an intellectual, a social and a practical activity down to the end of the nineteenth century. One lesson which seems to emerge is that science is not simply to be seen as a progressive activity. Every innovation, in theory, organisation, or instruments is double-edged. As more information is gathered, it becomes more difficult to see the wood for the trees; specialisation among scientists erects barriers, and can lead to the boring pursuit of 'normal science'; and technology can enrich or diminish life. Science is like any other human activity, and is not some unique and royal road to truth; the different sciences have had very different histories, and their practitioners have followed very different methods and had very different careers at various times and places. We must not expect to find in the history of science many general rules; the future will be different from the past, just as past pasts have differed from each other, and extrapolations are likely to be wrong. In our final chapter we must try to pull together what has been said so far, and to carry the story forward into the twentieth century.

8
EPILOGUE

In the twentieth century, as before, the boundaries of science have remained blurred. Disciplines such as archaeology have come to use techniques from physics and chemistry, and literary criticism to use computers; in literature, theories from psychology and metaphors from physics have been used as in the past; careers in science, and government spending on teaching and research, have expanded; and the transformation of society by technology has continued. It still seems appropriate to apply our pattern of science as an intellectual, a social and a practical activity even in our own period, when science has grown so much, and has come to receive more and more public attention.

In the late seventeenth century science emerged as a separate intellectual activity, distinct from theology, literature and history; and as a social activity with the first scientific societies. Publicists for the new philosophy urged its utility; but it was not until the end of the eighteenth century that many examples could be found of applied science. With the early nineteenth century came the development of sciences out of science; and this trend towards specialisation has continued apace into our own time, apparently inexorably and in spite of the unhappiness many men of science feel about it. Later in the nineteenth century came the formal courses in science at universities and other institutions; which have in the last hundred years proliferated, but still follow in the main the pattern laid down then.

In our own century, the characteristics of science have perhaps been its expensiveness—associated therefore with the need for governments or big businesses to support it—and its international character. Between about 1600 and 1900, science was predominantly a European affair, though there were important exceptions, and before 1600 it was otherwise. Our numerals had come from India by way of the Arabs; and much of the science that came from the Greeks had come to them from Mesopotamia or from Egypt. The Chinese seem to have contributed much to European technology. The system of piston, connecting rod, and crankshaft comes from them, as well as the double-acting principle for engines or pumps; while stirrups and horse collars, making possible the efficient use of horses for riding or for draught came to Europe from Asia in the medieval period. Islam not only passed on the Greek science to the West, but added to it—particularly in the realms of astronomy, chemistry and medicine. Paper, printing, and gunpowder came to Europe from the East; and it was not until well into the nineteenth century that steel made in Europe could compare with that of India or Japan.

When Europeans interested in the natural history of Asia made their collections, they found local artists ready to do splendid illustrations for them. The superb paintings made for John Reeves at Canton in the early nineteenth century are in the British Museum (Natural History), and a selection of them have recently been published. At much the same time, William Roxburgh at the Botanical Gardens at Calcutta was getting local artists to paint Indian plants for him; and these *Icones Roxburghianae* are also now being published. The exploration of the Himalayan regions was carried out in part by Europeans, but also largely by Indians trained to do the discreet kind of survey and writing up required of those travelling on behalf of the East India Company, in the midst of a suspicious population and in a remote area. In Peru too, illustrations of natural history were made for European naturalists by local Indian artists. Kaempfer's famous *History of Japan* was illustrated with Japanese plates of plants and animals; and there are fascinating Chinese works illustrating technical processes and devices.

Nevertheless, although science down to the twentieth century was clearly a cosmopolitan activity, it was the Europeans who from the late seventeenth century took the lead. They had the frameworks of theory, the vision of science as explanatory and useful, and the scientific societies, which brought modern science into being. It was therefore a European kind of science that was carried around the world, and the most prominent contributions to it during the eighteenth and nineteenth centuries were made by Europeans. Science is as much an expression of culture at a time and place as it is an approximation to truth about the world, and the science of the nineteenth century reflected the world view of those inhabiting the politically dominant continent. Even now, Japanese say that science still seems a foreign activity, not yet rooted in their culture. In Russia, to which Western science was transplanted over a century earlier, this is presumably no longer true; and certainly men of science from Asian countries, which now share the scientific institutions invented in Europe, have played an important part in science in the twentieth century.

We may expect that the science of the twenty-first century will be more fully international than that of previous centuries has been. Though the expense of some science will mean that some branches can be followed only in rich countries, or in international research centres, the most expensive science has not been in the past always the most interesting or the most useful. We may also expect that national or regional differences will remain between sciences as carried on in different parts of the world, as they have in Europe and North America; and that those who have made themselves familiar with more than one tradition or set of emphases, will be more likely to think of new ideas.

In the rise of science outside Europe, the most striking phenomenon in the twentieth century has been the transformation of the United States, from a provincial place in the world of pure science in 1900, to a position of leadership in many sciences fifty years later. We saw that in the nineteenth century there were many descriptive scientists in America, and that the sciences were liberally supported by the government, particularly if they promised to be useful. By 1900 America was technologically advanced compared to most European countries, and American universities were offering courses in the sciences; there were, moreover, by then a number of Americans who had an international reputation in the physical sciences. The growth since then can be put down in part to an exponential growth from these foundations, and in part to the wealth of the country. A major factor too was the exodus of men of science from Germany and Italy with the rise of fascism, which constitutes one of the great dispersals of learned men in history and is likely to be as important as any previous ones, such as the dispersal of scholars from Byzantium. Second only to the growth of science in the United States is that in the Soviet Union; and it is evidence of the powers of these two centres that English is now the chief language for the publication of science, with Russian next in importance.

In the twentieth century, then, there has been a vast expansion in the area where science is carried on; work of the first rank is now being done in places which were uncivilised or uninhabited only a hundred years ago. Science has, as we already noted, also become much more expensive. This applies particularly to the physical sciences. Faraday's apparatus for his work in chemistry and electricity was not expensive, and much of it he made himself—as his contemporaries did. But the coming of physical methods of analysis into chemistry, beginning with the spectroscope in the 1860s, brought into the chemical laboratory complicated and delicate pieces of apparatus that the chemist could not make. Companies also sprang up to provide him with glassware that he could make, but which was much less troublesome to buy. New but expensive devices have taken some of the drudgery out of chemistry, and perform analyses and other operations that would not have been possible in the nineteenth century. Thus fully discriminating between the 'rare earth' elements was a major problem at the end of the nineteenth century, and was only solved when Moseley in 1914 used the new technique of X-ray spectroscopy. To equip a chemical laboratory in the twentieth century is unavoidably more expensive than it was in the nineteenth century.

But it is in physics, and particularly nuclear physics, that the expense becomes enormous; and here too the isolated Dalton, Faraday, or Ampère has given place to a team of specialists who perform the various functions necessary to the experiment and the analysis of its results. This strikes some physicists as a new development, and a discontinuity or revolution within their science; but astronomy had been a 'big science' of this kind since the eighteenth century, and so had natural history with its teams of collectors transported in specially adapted ships, and its teams of taxonomists classifying the specimens sent home in the great museums. By the standards of the eighteenth and nineteenth centuries, astronomy, natural history, and earth sciences were very expensive, and could be carried on only with support from governments, great companies like the East India Companies, or unusually wealthy patrons.

We saw that in the nineteenth century there had been prestige projects, such as the exploration of the Arctic coastline of North America, the importance of which for geophysics, while not negligible, was small in proportion to the cost. On the other hand, governments were prepared to pay this cost and would not have paid smaller bills for probably more fundamental research. In this sense, the course of science was distorted by what the government would pay for; and in 1830 Charles Babbage wrote his indignant *Decline of Science in England* partly as a protest against the way money was so readily available for exploration, but not for mathematical physics in general or for his calculating machines in particular. It might have been that money put into his machines—which embodied some of the principles of our electronic computers—would have paid off better than the investment in exploration; but in fact the various coastlines were charted, while a clockwork computer would almost certainly have been impossible however much money had been made available to develop it. This we can say with benefit of hindsight; Babbage's contemporaries, faced with various pet projects all requiring funds, had to make the best choice they could, as administrators still do.

Exploration had the powerful backing of Barrow at the Admiralty, because it kept a part of the navy busy and in the public eye in peace time—though Barrow was also genuinely interested in geography. Connections of science with military activities again go back a long time, though they may have become more unpleasantly close during the twentieth century. Exploration of the Arctic or the moon by members of the armed services does not seem particularly sinister, compared for example to the development of rockets from the short-range affairs of the Napoleonic period down to those of today. For centuries men of science have been involved in the making and improvement of gunpowder; Lavoisier and Davy being two important chemists who worked on this problem. Lavoisier succeeded in improving the quality and consistency of French powder to a marked degree; but this involved searching people's cellars for saltpetre exudations, which did not endear him to the populace. Faraday was consulted during the Crimean War about a proposal to smoke the Russians out of their great naval base at Cronstedt with sulphur-filled fireships. Much earlier, the revolutionary and Napoleonic governments in France had mobilised scientists to solve problems connected with military defence and aggression. In Britain, Lord Brouncker, the first President of the Royal Society, made his reputation with papers on the recoil of guns; while in the nineteenth century as we saw it was in the armaments industry that mass production with interchangeable parts was first introduced.

Military involvment with science is not new, but it may well be that the change of scale—both in the numbers of those engaged, and in the nastiness of the new weapons designed— has brought about a new situation. In the early nineteenth century, new weapons could be applauded as a safeguard for civilisation against the incursions of some new goths or vandals; but we in the twentieth century have found that the goths and vandals are among us rather than beyond a frontier. As the wars of kings have become the wars of peoples, so attitudes among men of science have changed. The revolutionary and Napoleonic wars disrupted the international scientific community by making communications more difficult; but French and British men of science respected each other, and succeeded in finding out what those on the other side were up to. Thus Davy and Guy-Lussac heard rapidly about each other's experiments on potassium and on chlorine; and it was Berzelius in Sweden who found it difficult to get news of their work. A century later there was, during the First World War, a real hatred of German scientists in Britain, although most leading chemists had done their research in Germany and had usually enjoyed it. German scientists were expelled from foreign membership of scientific societies, being held ultimately responsible for real or alleged hunnish atrocities; and no doubt the odium was returned from the other side.

What had been patriotic in the early nineteenth century— that is to put your knowledge at the disposal of your country to at least some degree—had by the early twentieth century become disgraceful, to those on the other side. War had ceased to be an intelligible competition for territory, trade or power, and had become some kind of crusade or fight to the death.

Since in such a war all is fair, men of science did become involved in work with poison gases, bombs, torpedoes and so on. The end of this process was seen in the incendiary bombs, rockets and atom bombs of the Second World War and the bigger and better versions of them made since. These developments have made even compatriots of the scientists involved ask whether such work is patriotic or immoral. The change in scale of weapons and wars had brought about a change in attitude towards the morality of science, and exposed the ambivalence of science. It has been said that scientists take credit for penicillin, but put the blame for the atom bomb on to generals or politicians. There is too much truth in this for comfort; though it would seem that nowadays there are so many vociferous critics of science and technology that the debit side of the balance is alone remembered.

Military involvement also brought into science a new feature, secrecy. This has always been associated with technology, but one of the hopes of the natural philosophers of the late seventeenth century was that the scientific societies would bring craft secrets out into the open. Certainly openness has always been associated with the sciences. The results of experiments should be published so that they become public knowledge, and can be repeated; or so scientists have always believed. There have been hoaxes, like Piltdown Man, in the past; and curious episodes, as when W. H. Wollaston encouraged Richard Chenevix in 1803 to publish an analysis of palladium which he knew to be wrong, in order it seems to make a fool of him. There have also been a few scientists who have never got around to writing up their work in publishable form; of whom the most notable might be Henry Cavendish and Joseph Banks. These were both aristocrats who flourished about 1800, and their case must be unusual; in our century at least, those who publish papers although they have no interesting work to report are much more plentiful. But at all events, the norm was recognised that papers should be a full and honest account of work done; and on this candour and openness the cumulative vision of science was based. Work done in military establishments is not public knowledge; and the existence of such work involves boundaries within the scientific community that are new and must often be galling and wasteful.

Science in our century is threatened, or so it is said, not simply by soldiers and their governments but by a military-industrial complex. The idea that discoveries should be kept secret has long been associated with industries; and the scientist working in industry gets his satisfaction from helping to make a better or cheaper product rather than from seeing his name in print. The involvment of men of science in industry is again something that is far from new. But it is also something which has grown in scale with the development since the nineteenth century of industries which are more closely based upon science than older industries were. The demands of industry are not only for secrecy rather than openness, but also usually for a short-term usefulness. This leads to demands from industrialists for more practical courses in universities; and for their purposes they may be right to urge this, for the close connection which publicists for science saw between pure research and economic development seems to be illusory. As we found in looking at the nineteenth century, some of the most important discoveries that led to technical advances were only developed into technology a generation or more after they were first made known—the great example of this being the electric dynamo. The technological development may well then take place in a different country from that in which the scientific discovery was made. Technological innovation and scientific discovery are two different and only loosely connected affairs.

The later twentieth century has seen a certain disillusionment with technology, which is identified with a rapacious attitude towards the resources of the world and a selfish determination to keep rich countries rich. Here one must surely remember that technology is no more monolithic than is science. Just as there is no unique scientific method, attitude, or style of theory and experiment, so there is no single approach in technology. Different engineers, such as the sober Stephenson and the flamboyant Brunel, have produced different kinds of works even though they were engaged in the same field, and there is little in common between all those engaged in the various technologies except that they are performing a practical job and are subject as a rule to constraints on time and cost. Just as in the sciences the solving of one problem raises some more, so it is in technology; nobody can see into the future, and practical decisions have to be taken on the best evidence available.

It does not detract from the achievement of those engineers of the nineteenth century who provided sewerage systems for the great cities, and thereby got rid of cholera, typhus and typhoid that, by the twentieth century, the sewage was polluting the sea; just as it is no slur on physics or on Newton to say that his treatment of space and time made difficulties for his successors in the nineteenth century. We live in the kind of world in which all our problems will never be solved, and in which all life is an achievement against the odds—the laws of thermodynamics will entail the eventual death of us all, and eventually too the 'death' of the solar system. Whether it is the best of all possible worlds is an open question, but certainly a world without problems would be duller than ours.

The problems of urban living with which the engineers, architects and medical men of the nineteenth century had to wrestle were themselves the effects of technology. The dark satanic mills of the eighteenth century, making cotton, ceramics or steam engines gathered around themselves cheaply-built towns, with sanitation appropriate to country villages. The perfection of the water closet flooded the rivers with sewage which had previously remained each householder's problem in his cess-pit. Canals, turnpike roads, and railways made possible the supply of goods to the rapidly growing cities; and could eventually be used to bring in fresh and unadulterated food, while new sewers carried off waste, and reservoirs and aqueducts brought in filtered and chlorinated water. These in turn made possible the further growth of the city, and then the movement of people into suburbs, leading to fresh problems in city centres. The effects of one technological revolution could only be dealt with by further and different technological developments, which in their turn had unforeseen consequences with which we have to wrestle.

No one, of course, wants to encourage irresponsible innovation: of which an example might be the famous chemical plant at Minamata in Japan where many people were poisoned over a number of years by mercury compounds discharged into the sea and absorbed by fishes which were then caught and eaten. This chemical company was apparently keen on its reputation for rapid development of processes up to the industrial scale, which can be profitable as well as prestigious; and the factory was built in a relatively backward area economically, so that the whole enterprise seemed to be in the national interest. At first enquirers into the mysterious and horrible deaths in the region did not think of mercury poisoning; and when they did it seemed at first incredible that so valuable a substance could be allowed to escape into the sea. When the case was made out, the authorities were reluctant to act firmly; and it seems that the poisoning only stopped because the process which a few years before had been an innovation became obsolete with the rise of the petrochemical industry. There are and always have been alarming stories of this kind, especially where there has been an ambitious technical advance; the spires of most medieval churches have fallen down at least once, bridges and dams have always been liable to collapse, ships sank, and railway trains collided. Not every contingency can be allowed for, and all living involves taking risks; we cannot ask for a completely safe technology, but we can hope for a responsible one, where known risks are not ignored.

It seems, then, as though the new characteristic of science in the twentieth century is its size. It is carried on in more countries by more people at more expense than ever before; and similarly, changes in technology affect the lives of more people than they have in the past. Whether on the other hand those who live in Western Europe or North America have gone through a period of more profound technical change than their great grandfathers did, with gaslight, railways, telegraphs and cheap postage, is an open question. Certainly it seems that even earlier our ancestors in the Romantic Period—twenty years or so either side of 1800—felt the same malaise about science and technology. Blake, Goethe, Schelling, Hegel, Coleridge, and Keats —of whom most knew a good deal about science—were unhappy about the view that the world was a kind of vast mechanism, and man no more than a smaller mechanism; and that nature could be raped for man's benefit. They favoured a biological view and hoped for a dynamic science, in which the apparently static was really in equilibrium between opposed forces. Nothing in the world persisted except as people, waterfalls and columns of smoke persist, through the ceaseless flux of their material particles. Such a biological and dynamical view is akin to that held today by those who see ecology rather than physics as the most important science, revealing what really goes on in the world: a balance of nature, in which an apparently small interference can produce a whole series of changes throughout the whole system.

When we apply our framework to the science of the twentieth century, it still seems to fit despite the change in size and the associated change in pace. Because more people have been involved in science, it has undergone more rapid change—and possibly progress—than before; and the process of specialisation has continued apace. The tensions between those pursuing the various sciences have not disappeared; and the frontiers between the various sciences have remained fluid—parts of geology have been taken over by physics as geophysics, and parts of botany by chemistry as biochemistry. That is, questions which a generation or so ago would have been answered simply in geological or botanical terms are now answered in terms of physical or chemical mechanisms which seem more general and fundamental. This is the kind of process which we found happening in the nineteenth century and before. The process has not been all one way; field naturalists for example have fought back, and most biologists would no doubt concede that a biochemical view must be supplemented by an ecological one. The reduction of all the processes in the world to physical and chemical terms is, as it was in the days of Descartes, no more than a goal attractive to some.

It is sometimes supposed that whereas the view of the history of science as a series of revolutions separated by long periods of steady normal science—a view proposed by Thomas Kuhn—fits early science quite well, it cannot be applied to that of our own day, which is to be seen as a steady advance towards truth. This view supposes that we now have the proper concepts, and that revolutionary changes of view are no longer necessary or possible. Kuhn's paradigm is attractive, because it treats science as an expression of the culture of a time and place and not as a steady march towards truth—though this element is distasteful to many scientists, who dislike thinking of their activity as a kind of intellectual game in which the rules are changed every so often when play seems to be getting bogged down. It can be criticised because Kuhn seems to look almost entirely at factors within science; and because it seems impossible, once one looks closely at any part of the history of science, to say which changes are revolutionary. It is too like the old-fashioned view of history in which all changes have been brought about by a few heroes—an account which always seems more plausible for the past than the present.

Be that as it may, Kuhn's paradigm can be applied to episodes in twentieth-century science just as well as to those in the more remote past. This has in fact been done for the revolution in the earth sciences associated with the theory of continental drift, and its later development into plate tectonics. Tension between physicists and geologists became obvious when in the later nineteenth century Lord Kelvin tried to apply the Second Law of Thermodynamics to the cooling of the earth and the combustion of the sun in order to find out how old the earth and the solar system might be. His figures were low—less than 100 million years—and did not seem to give long enough for the geological and evolutionary processes in which most geologists, following Lyell and Darwin, believed. There was some consternation among geologists, and some bluster from T. H. Huxley; but it was not until the early twentieth century and the discovery of radioactivity, that a source of engery in both the earth and the sun was found of which Lord Kelvin had known nothing and which completely upset his calculations. It turned out that geologists had been right to ignore the theories of physicists; and many generalised from this one case, and assumed the autonomy of their science.

They also seem to have eschewed wide-ranging theorising in favour of Kuhnian normal science; and thus the early speculations about continental drift were ignored or deplored by the majority of geologists down to the 1960s. By then Arthur Holmes in Britain and Tuzo Wilson in Canada among others had begun to produce evidence for continental drift, and a plausible mechanism for it. These both depended upon geophysical studies, of magnetism and of the behaviour of magmas; and over about a year the majority of geologists changed their view of what was to count as evidence in their science, and accepted the new theory as well. This change of viewpoint seems to fit well Kuhn's scheme, constituting a revolution rather than a predictable advance in normal science.

Another example of a revolution in recent years might be the postulation of the double helix structure for DNA by Crick and Watson. This involved the reinterpretation of evidence from fields such as crystallography, chemistry, and genetics which had previously seemed almost unrelated; some of the evidence was new, but more important was that the facts were seen in a new light. If normal science is the province of questions difficult to answer, and revolutionary science that of questions difficult to ask, then once again such a step seems revolutionary. In the 1960s as before, there were times when the existing concepts, conventions, and academic frontiers seemed to block further progress, and some more radical change of viewpoint was necessary. Science today as in the past is not simply a collection of definite facts; conceptions are superimposed upon them, and it is as important to get hold of appropriate ideas as of well-established facts.

Some revolutions indeed make the facts which a previous generation had painstakingly established look of little interest. Thus Moseley in 1914 demonstrated to the satisfaction of most contemporaries that it was the atomic number, or positive charge on the nucleus, that characterised a chemical element. Since Dalton and Berzelius a hundred years before, the relative atomic weight had been accepted as this characteristic feature; this is clearly a much less theory-loaded criterion, but it had by 1900 led to some anomalies—tellurium and iodine seemed to come in the wrong places in Mendeleev's Periodic Classification of elements, based upon atomic weights. But many eminent chemists spent much of their working lives establishing atomic weights to ever higher degrees of accuracy, believing that these were one of the constants of nature, and would enable chemistry to be made into a mathematical science. Moseley's theory, developed by others after his death at Gallipoli in 1915, made atomic weights almost irrelevant; the atoms of an element might be of various weights—those of different weights were by Aston later called isotopes—and the true constant of nature was the atomic number. The same element from different places might, if it were composed of a different mixture of isotopes, have different atomic weights; fortunately for chemists this is not common except among the radioactive elements, but it was found to be the case for lead. What had seemed to be a fundamental quantity now became an adventitious one; in the march of science, facts may cease to be brute facts and become interesting, and then perhaps with another change of viewpoint, become brute facts again.

The kind of analysis of science as explanation which fitted it in the past seems thus still applicable; science is an ongoing process, involving creative thinking as well as careful experimentation. These two activities are not as separated as Kuhn's analysis might imply; in the history of science as in the history of the earth, changes have been brought about by the slow operation of causes now at work, and not by series of cataclysms —the revolutions in the history of science are no more than periods of unusual activity and upset in some field. As an intellectual activity in relation to other activities, science has again in the twentieth gone on much as before.

The process whereby men of science began to take upon themselves the role of wise men or prophets, who can advise their contemporaries on all kinds of topics, has continued from the nineteenth century when men like T. H. Huxley took over this role from the clergy. In the biological sciences, there is no lack of Malthusian prophets proclaiming that the end of the world is at hand; but in other regions of science, failures in predictions—of the costs of Concorde or of nuclear power, for example—and such scandals as the spread of pollution and the invention of horrible weapons, have weakened the practical and moral authority of the prophets. There seems on past showing to be little reason to believe that a world governed by scientists would be a better place. There is a room for experts—one would not want to be put into the hands of an amateur heart surgeon—but no reason to put them in charge; there is no need for a professional surgeon to run the National Health Service.

Pollution should be a problem for scientists to solve rather than a stick with which to beat scientists; the bad smell does not really indicate corruption in the world of science, although it is often taken that way. But if the moral and prophetic authority of scientists has in the last ten years or so become weaker than it was in the earlier part of our century, the authority of scientific explanation as a paradigm of explanation seems to survive as healthily as ever. Philosophers and social scientists seem in the twentieth century, as before, to have often taken as the paradigm of explanation that in vogue in parts of physics; and to see all other kinds of explanations, even in the sciences, as better or worse approximations to it.

In fact it seems that, as we found for past science, there are across the sciences a range of kinds of explanations; and indeed in twentieth-century physics there have been considerable upsets because the concept of causality which had been inherited from the days of Galileo no longer seemed applicable to the world of atoms. Physicists in our century have been less certain than their admirers that they know how to explain anything.

In the later nineteenth century, metaphors from Darwin's theory of evolution by natural selection were taken up by social scientists and politicians, and used to legitimise or attack a *laissez-faire* society. Similarly in the twentieth century, Einstein's conception of space and time as relative rather than absolute has provided a metaphor for those who urge the relativity of moral precepts; and Bohr's 'complementarity' has been a generally useful idea. By the end of the nineteenth century, professional biologists were feeling unhappy about Darwin's theory, in the light of current work on genetics; but by then the theory had gained a firm hold upon the public mind, and was generally believed to be established. Something similar seems to have happened in our day with the theories of Freud in psychology; which are accepted only by a sect among psychologists and even psychiatrists, but are widely taken for granted among laymen so that Freud's terminology has even passed into ordinary language as 'survival of the fittest' and 'struggle for existence' did.

As science has spread beyond the bounds of Christendom in this century, so one might expect that the relations between it and theology would have become less close; and this indeed seems to have happened, particularly as in the West the religious revival of the nineteenth century has died down, and the churches have declined from their position of enormous importance. Some aspects of science have still attracted the attention of theologians; who have seen in the insights of scientists an element of 'cosmic disclosure' as well as seeing in military and industrial science a number of moral questions. In our day as earlier the creation of the world has exercised the imagination of men of science, with controversy between those who suppose it began at some moment of time—at present this view seems to be in the ascendant—and those who like Aristotle and Hutton adhere to a steady-state view, finding 'no vestige of a beginning, —no prospect of an end'. The former conception accords with the notion of a Divine creative *fiat*, while the latter draws attention to the Divine sustaining of the created order; neither view confirms or overthrows any religious understanding of the world, but both are congenial to certain theological positions and inimical to others. Among men of science, the older natural theology with its arguments from design to a Designer, still retains its hold in some quarters; while psychologists and sociologists have made phenomenonological studies of religion in the wake of William James and Max Weber.

The old hostility demonstrated in the Huxley-Wilberforce debate of 1860 has died down; but one of the issues in that debate, the unity of mankind, is still alive. Bishop Wilberforce was the son of one of those most responsible for the ending of the slave trade, who believed that all mankind was one whatever their colour; whereas some early evolutionists about 1800 saw a steady progression from orang-utans through blacks to whites, which might give the superior race the right or the duty to rule or enslave the inferior. Evangelical anthropologists such as J. C. Prichard in the first half of the nineteenth century laboured to show that all mankind was one species, and was completely distinct from the apes. Richard Owen, the great palaeontologist and Director of the British Museum (Natural History) when the *Origin of Species* came out, had supposed that Adam and Eve were probably black; and Prichard thought that whiteness was a result of civilisation.

The idea that apes were like men had been familiar since antiquity, for Galen when unable to get human corpses for dissection had had to make do with those of Barbary apes. In the seventeenth century, Edward Tyson had published an account of the anatomy of an anthropoid ape; but it was not until about the end of the eighteenth century that the offensive idea gained ground that species were unreal, nature a continuum, and some peoples more like apes than others. It was this which Owen thought he saw in Huxley's work which was eventually published as *Man's Place in Nature*; and his friend

Wilberforce spotted in the *Origin of Species* passages about slave-making ants to which he drew attention in his review, which seemed to imply that it was natural to enslave, and even natural that it should be black ants or men who should be made slaves. Darwin himself had seen slavery in Brazil, and had very firmly in public and in private expressed his revulsion at it; but Darwinism did in the event seem to provide some support for racialism and for ruthless social policies, and we ought to be careful about deriding those who opposed it. It may be that their hearts were in the right place.

Thus to suggest that all mankind was descended from apes bore harder on some men than on others, for they seemed less remotely descended. In the twentieth century, the history of mankind has been pushed back by discoveries of fossils from a time long before the Victorians suspected that man existed. Men and orang-utans would be generally agreed to have a common ancestor, but only a very long time ago; and there can be few now who would assert that the different races of man are different species or even different subspecies—that all men are interfertile has been long known, indeed ever since sailors undertook long voyages. But there are those today who assert that the genetic differences between the various races are not limited to superficial characters like colour of skin and hair, but also include intelligence. Clearly if some races of mankind are naturally stupid, they cannot all be treated in the same way; and most scientists have shown their moral repugnance at this notion, as Owen and Wilberforce did, by refusing to look seriously at the alleged evidence for it. To do this is not necessarily to be unscientific; since the late seventeenth century most men of science have not felt it necessary to investigate ghost stories seriously, or since the late eighteenth century to look hard at proposed perpetual motion machines, because in their world view such things are impossible.

The scientific community is thus exposed to tensions very like those to which it was subject in the past. To the outsider the gaggle of absent-minded professors at whom the wits of about 1700 could laugh, has been transmuted into an alarming collection of dangerous and probably amoral sorcerers—which was how medieval natural philosophers were often seen too. But from the inside the differences are mostly those connected with size, and with the increase in formality which that brings. The sciences have become increasingly fragmented as specialisation has gone on rapidly; old journals have become more specialised, and new specialised ones have been founded. In the nineteenth century, men of science in one country might not hear about the work of those in another because they never came across the journals in which their work was published. This problem was reduced with the rise of abstracting and translating journals; but by the twentieth century, the problem became one of finding in the mass of published material something that one wanted to know.

Journals proliferated at such a rate by the middle of our century that it became impossible to read all that was published in any but a very narrow field; and papers in a journal unfamiliar to a given scientist, which were abstracted under a heading that did not seem to him relevant to his work, would never come to his attention although they could save him unnecessary labour. The nineteenth-century problem of a man repeating work because he did not know it had been done already has not yet been solved. This is different from the phenomenon of simultaneous discovery, or the races to synthesise a compound or resolve a structure which are a characteristic of relatively normal science; in a race, the competitors usually know who they are up against, as we find in the story of how the structure of DNA was worked out. With the 1970s has come a further problem, that of finding the money to keep up subscriptions to scientific journals; so that libraries have had to cancel important journals of which they have long runs because of the possibly short-term consideration that other journals are at the moment closer to the research interests of those who happen to be in the physics or chemistry department.

To keep abreast of the field is no easier than it was a hundred years ago, and so much is written that if anybody were to try to read it all he would have no time for any research himself. One way round this problem is the conference or congress; and international congresses have grown in importance and frequency since the first really important and famous one, on chemistry, was held at Karlsruhe in 1860. Conferences are not cheap either; but they do keep the scientific community together as the scientific societies of the seventeenth century did when it was much smaller, and covered much less of the world. Sometimes the papers read at such congresses are important and interesting enough to make good reading later, when the *Proceedings* are published as they all too often are; but usually it is the less formal aspects of the conference that are more important, because it is a place to meet face to face those whom one has previously only encountered in their writings. The printed reports of such affairs are thus likely to be chiefly of value to the social historian of science in the next century, who may want to known who was chosen to be chairman of a session, and who gave the keynote address—though one may hope that he will have some secret histories to supplement the official accounts.

Part of the problem of keeping up in the twentieth century is connected with the way in which science as a career has developed. As is well known, the dictum 'publish or perish' is all too frequently acted upon. The idea of open science developed with the journals of the seventeenth century was that anybody with anything interesting to report should send it to the editor without delay. Even then some of the papers seem to have been sent in by those who were anxious to see their names in print even though they had nothing much to report; this is a phenomenon which will be familiar to editors of parish magazines as well as learned journals. With the early nineteenth century the scientific paper began to be used as a measure of a man's capacity and energy; so that in the discussions about 1830 about the nature of the Royal Society one of the complaints was that many of those in high places in the world of science had published little or nothing. Publication of papers is easy to quantify; one counts number of papers, or number of words, and thus one obtains an estimate of merit.

Some of those who had not published were indeed amateurs in the modern and rather derogatory sense; that is, rather than being informed critics, they were dabblers. Others were informed critics and patrons, or amateurs in the eighteenth-century sense; while others again were gentlemen like Cavendish and Banks, whose status in the small world of science was secure without their having to publish their work in a formal manner. Davy, making his way in the social and scientific worlds, published a great deal by the standards of the day; and so in France in the eighteenth century had D'Alembert, who had also to make his way from humble origins. To contemporaries and to us, it was clear that many of the papers of these two were provisional, not to say half-baked, in character; they reported work in progress, in order to keep their name in the public eye and to stake out claims to a territory. This can sometimes be a useful procedure in a fast-developing branch of science, as mathematical analysis was in the eighteenth century and electrochemistry was in the early nineteenth; but it can also lead to rambling papers, in which little of real interest is presented in an ill-digested form.

In the last hundred and fifty years, the number of scientists like Davy and D'Alembert in this respect has grown enormously; and there are correspondingly fewer like Cavendish, Banks, Wollaston, or Young whose reputation is secure and who can wait until they have finished a course of research to their own satisfaction before publishing an account of it. Personal recommendations still play an important part in advancing the career of a young scientist, and conferences—particularly national rather than international ones—have sometimes a disagreeable resemblance to a hiring fair. But more often than not one comes across somebody through reading a paper he has written before one meets him. The small world of nineteenth-century science, with its informal structure, has given way to a larger and more formal world in which to publish a paper is the way to attract and keep attention, and thus to advance one's career. The number of half-baked papers has increased much more rapidly therefore than the number of polished papers; and the paper as a means of self-publicity has almost overtaken the paper reporting some new and interesting investigation—or so it can easily seem to the jaundiced.

Thus, as one might expect, as science has become increasingly a career the possibilities of careerism have greatly increased. The man who publishes a steady stream of papers may be better known, and therefore be more rapidly promoted, than another who writes less but whose papers are well thought out and based upon carefully assessed data. Those who publish a lot tend to be repetitious; and one may suspect that when great series of experiments are described, many of them have been performed entirely by research assistants or graduate students.

This phenomenon is not new; there have often in science as in the arts been times when the most original thinkers—as seen from a later perspective—have languished while more common-place people have forced themselves into the public eye and have occupied the important offices. Thomas Thomson in the early nineteenth century got most of his research on atomic weights done for him by research students, and later on Crookes and Lockyer got much of theirs done by paid assistants. All these three were editors of journals; they were busy men trying to work their way to the top of the tree in science, which in the event none of them quite did.

Patronage in science opens a promising field to the historian and the sociologist, who are now beginning to dig in it. Men like Pallas and Banks in the eighteenth century could provide it through their position in the world of science in Russia and in Britain; but there was nothing surprising in that, in a world which worked through patronage. By the nineteenth century we begin to find a more academic kind of patronage, where the prestige of a teacher is such that he is invited, or at least is able, to place his pupils in such good jobs as come up. Linnaeus in the late eighteenth century had been in this position; Berzelius in the early nineteenth century, and then Liebig, enjoyed this commanding role; and in our century Rutherford at the Cavendish Laboratory had a grip on physics of this kind between the wars. One's reaction is to see this as a disease in science, like the excessive publication by those in fear of perishing; but just as there can be a value to others in a premature or provisional paper, so it may be that the best students are in fact attracted to the best teacher, whose graduates therefore are in their turn the best teachers of their generation, and can spread in other previously benighted institutions the best practice which they have learned from the great man.

The process will anyway in the long run be self-correcting, as the pupils of the great man themselves organise research schools and start placing their own students; and this for example happened in nineteenth-century Germany where Liebig's methods of practical teaching were rapidly disseminated by his students. It is also something that will perhaps become of less importance as the scientific community grows, because there will no longer be just one outstanding centre in any one discipline—though as science becomes fragmented, it is still possible for one place to be the focus of exciting developments in one field, though this would be narrower than it might have been a hundred years ago. Some kind of patronage is bound to survive, and one may expect that the pattern will become more complex.

Those making a career in academic chemistry in Victorian Britain were generally expected to have done some research in Germany; and it may be that in the later twentieth century again it will be valuable for a scientist's career that he has spent some time abroad. Jobs and promotions in developing countries often at present go to those who have worked abroad, and have got up their foreign languages so that they can read papers in European and North American periodicals; but one could argue from the history of science that foreign experience would be useful too for those in scientifically developed countries. Indeed there are now a fair number of fellowships available for such work; and if nineteenth-century experience is anything to go by, then work in a different country involves work in a rather different tradition, and thus a greater probability of gaining a new insight.

Because even in the most prosperous times and countries, money is limited and so are positions of importance in the world of science, the growth of any one branch is bound to be more or less at the expense of others. This Darwinian situation is naturally most evident in periods of stringency; just as nature is most obviously red in tooth and claw when food is short. Science nowadays is no more monolithic than it ever was; its boundaries are as uncertain as ever; and there are situations where the interests of engineers, chemists, mathematicians, geographers and psychologists are one, and others where they are opposed. Neither as an intellectual activity nor as a social one are the boundaries of science clear and distinct; the frontiers between mathematics and logic, or between psychology and sociology, are as conventional as those between physics and chemistry.

In discussing the characteristics of twentieth-century science we have already focused upon its relationships with technology and with governments, and noticed the differences which the great increase in scale of science as a practical activity has brought about. Radical critics of science see pollution and armaments and social injustice as its practical fruits; see explanation as merely a prelude to application; and see no more than careerism and nest-feathering in the professionalisation of the sciences. Our discussion has indicated that the situation is—like any real situation—much more complex and interesting than that; and that the nature and role of the sciences is something that is worth understanding.

This understanding will only be tolerably complete if it has an historical basis, for the sciences have grown through time and while there is much in them that is new in our century, there is much more which has survived from the past, and all the important aspects are found at least in embryo there. If we are to gain this understanding through history, we must not be too preoccupied with scientific progress; it is change rather than progress that we must investigate. Science is an expression of culture, and its manifestation in different times and places has been different and will no doubt continue to be so. It is also complicated; and unless it is seen as an intellectual, social, and practical activity it cannot be seen whole. It is then one of the most fascinating and rewarding of human activities to study; and surprisingly it is one which so far has been relatively neglected, and left to specialists. Science is too wide-ranging and important an activity to be left to experts, and the same is true of its history. It is there that we shall find science as an ongoing process of coming to grips with nature, rather than just the accumulation of millions of facts; and find that it is as much a matter of asking questions as of supplying answers. Its history, as it is written, should have the same character.

SUGGESTED FURTHER READING

1. T. S. Kuhn's view of science can be found in his *Structure of Scientific Revolutions*, 2nd ed., Chicago, 1970; a brief early abstract of it appears in a useful volume. A. C. Crombie (ed.), *Scientific Change*, London, 1963. With this view, we may contrast that of C. C. Gillispie, who in his *Edge of Objectivity*, Princeton, 1960, provides a very readable account of scientific progress. A Marxist perspective may be found in J. D. Bernal's *Science in History*, 4th ed., London, 1969; while interactions between science and technology are also stressed in D. S. L. Cardwell, *Technology, Science and History*, London, 1972. Karl Popper's view of science as a matter of attempts to falsify hypotheses can be found in his *Conjectures and Refutations*, London, 1963. Newton as one of the last of the magi can be found in F. E. Manuel, *A Portrait of Isaac Newton*; while the experimental Newton who provided a paradigm for Franklin is in I. B. Cohen, *Franklin and Newton*, Philadelphia, 1956 – see also his introduction to the variorum edition of Newton's *Principia*, Cambridge, 1971. For a discussion of the use of historical evidence in the sciences, see my *Sources for the History of Science*, London, 1975; and for booklists and discussion, my *Natural Science Books in English*, London, 1972.

2. On Harvey, see W. Pagel, *William Harvey's Biological Ideas*, Basel, 1967; and contrast with this the more 'positive' accounts in G. Keynes, *William Harvey*, Oxford, 1966, and G. Whitteridge, *William Harvey and the Circulation of the Blood*, London, 1971. Pagel attaches much more importance to Hermetic ideas, on which see also A. Debus, *The English Paracelsians*, London, 1965, and his forthcoming *History of Chemical Philosophy*; and K. Thomas' fascinating *Religion and the Decline of Magic*, London, 1971. On the Copernican theory, see T. S. Kuhn, *The Copernican Revolution*, Cambridge, Mass., 1957; and A. Koestler's *The Sleepwalkers*, London, 1959. On the 'reduction' of biology to physical science, see E. Mendelsohn, *Heat and Life*, Cambridge, Mass., 1964; and for superb essays on the history of palaeontology, M. Rudwick, *The Meaning of Fossils,* London, 1972.

3. On natural theology, see H. R. McAdoo, *The Spirit of Anglicanism*, London, 1965, and O. Chadwick, *The Victorian Church*, London, 1966–70; and the materials published for the Open University course on Science and Belief. On magic, see the writings of Frances Yates, and especially her *Giordano Bruno and the Hermetic Traditon*, London, 1964; on Galileo's trial, G. de Santillana, *The Crime of Galileo*, London, 1958; and on the shift of power and prosperity to the north of Europe, R. Davis, *The Rise of the Atlantic Economies*, London, 1973. On Darwinism, see C. C. Gillispie, *Genesis and Geology*, Cambridge, Mass., 1951; L. Eiseley, *Darwin's Century*, London, 1959; and D. L. Hull, *Darwin and his Critics*, Cambridge, Mass., 1973. On science and the fine arts, see F. D. Klingender, *Art and the Industrial Revolution*, ed. A. Elton, London, 1968; G. Grigson's ed. of Thornton's *Temple of Flora*, London, 1951; and M. Twyman, *Lithography, 1800–1850*, London, 1970. On the spirits, A. Gauld, *The Founders of Psychical Research,* London, 1968.

4. On the rate of growth of science, see the notes to chapter 8. On scientific societies, see W. E. K. Middleton, *The Experimenters: a Study of the Accademia del Cimento*, Baltimore, 1972; M. Purver, *The Royal Society*, London, 1967; R. Hahn, *The Anatomy of a Scientific Institution: The Paris Academy of Sciences, 1666–1803*, Berkeley, 1971; M. P. Crosland, *The Society of Arcueil*, London, 1967. On different national traditions, see the useful collection of papers in M. P. Crosland (ed.), *The Emergence of Science in Western Europe*, London, 1975; C. Russell, *The History of Valency*, Leicester, 1971, and W. L. Scott, *The Conflict between Atomism and Conservation Theory*, London, 1970, also bring out international differences in theorising. On Dalton, see A. Thackray, *John Dalton*, Cambridge, Mass., 1972; on science in America, N. Reingold, *Science in Nineteenth-century America*, London, 1966; on Banks and the Royal Society, A. M. Lysaght, *Joseph Banks in Newfoundland and Labrador*, London, 1971; and W. R. Dawson, *The Banks Letters*, London, 1958. On Linnaeus, see F. A. Stafleu, *Linnaeus and the Linneans*, Utrecht, 1971; and W. Blunt, *The Compleat Naturalist*, London, 1971. On Pallas, see U. Urness (ed.), *A Naturalist in Russia*, Minneapolis, 1967. On late Victorian Science, J. Meadows, *Science and Controversy; a biography of Sir Norman Lockyer,* London 1972; on a professional scientific society, R. C. Chirnside and J. H. Hamence, *The Practising Chemists*, London, 1974—more ambitious histories of the Royal Institute of Chemistry and of the Royal Institution are being

prepared, under C. Russell and F. Greenaway respectively. For science in one country, see J. Challinor, *The History of British Geology: a bibliographical study*, Newton Abbot, 1971.

5. On the founding of the British Museum, see G. R. de Beer, *Sir Hans Sloane and the British Museum*, London, 1953. On optical instruments and technical frontiers, S. Bradbury and G. L. E. Turner, *Historical Aspects of Microscopy*, Cambridge, 1967. On the Franklin Institute, B. Sinclair, *Philadelphia's Philosopher Mechanics*, Baltimore, 1974; on botanical illustration, W. Blunt, *The Art of Botanical Illustration*, London 1950, and the books of A. Coats, *The Book of Flowers*, London, 1973, and *The Treasury of Flowers*, London, 1975. B. Henrey, *British Botanical and Horticultural Literature before 1800*, 3 vols., Oxford, 1976. G. Caley, *Reflections on the Colony of New South Wales*, ed. J. E. B. Currey, London 1967. I am preparing a work on zoological illustration. C. Close, *The Early Years of the Ordnance Survey*, 2nd ed., ed. J. B. Harley, Newton Abbot, 1969; H. Quill, *John Harrison*, London, 1966. M. Daumas, *Scientific Instruments of the 17th and 18th Centuries and their Makers*, tr. M. Holbrook, London, 1972; H. Michel, *Scientific Instruments in Art and History*, tr. R. E. W. and F. R. Maddison, London, 1967; R. A. Salaman, *A Dictionary of Tools*, London, 1975; W. Steeds, *A History of Machine Tools*, Oxford, 1969. On the teaching of science, T. J. N. Hilken, *Engineering at Cambridge, 1783–1965*, Cambridge, 1967; M. Travers, *Sir William Ramsay*, London, 1956 —on the first grants for science at universities, among other things; G. W. Roderick and M. D. Stephens, *Scientific and Technical Education in 19th Century England*, Newton Abbot, 1973.

6. H. Woolf, *The Transits of Venus*, Princeton, 1959; C. Lloyd, *Mr Barrow of the Admiralty*, London, 1970; J. C. Beaglehole (ed.), *The Journals of Captain James Cook*, 4 vols in 5, Cambridge, 1955–74; J. F. W. Herschel (ed.) *A Manual of Scientific Enquiry*, 1851 ed. reprinted with introduction by me, London, 1974; M. Deacon, *Scientists and the Sea*, London, 1971; A. Day, *The Admiralty Hydrographic Service, 1795–1919*, London, 1967; G. S. Ritchie, *The Admiralty Chart*, London, 1967; G. Williams (ed.), *Documents relating to Anson's Voyage round the World, 1740–44*, London, 1967; A. E. Gunther, *A Century of Zoology at the British Museum through the Lives of Two Keepers, 1815–1914*, London, 1975—the Keepers were J. E. Gray and A. Gunther. E. J. Russell, *A History of Agricultural Science in Great Britain*, London, 1966.

N. Rosenberg (ed.), *The American System of Manufactures*, Edinburgh, 1969.

7. D. S. L. Cardwell, *From Watt to Clausius*, London, 1971, is a very useful history of thermodynamics in its theoretical and practical aspects; and on the Industrial Revolution generally, see D. Landes, *Unbound Prometheus*, Cambridge, 1969. C. C. Gillispie (ed.), *A Diderot Pictorial Encyclopedia of Trades and Industry*, 2 vols., New York, 1959; M. Faraday, *Chemical Manipulation*, 1827 ed. reprinted with introduction by G. Porter, London, 1974; L. P. Williams, *Michael Faraday*, London 1965—and on Faraday, T. Levere, *Affinity and Matter*, Oxford, 1971. F. Szabadaváry, *History of Analytical Chemistry*, tr. V. Svehla, Oxford, 1966, is interesting on apparatus. V. Ronchi *The Nature of Light*, London, 1970, is a history of optics. R. E. Schofield, *The Lunar Society of Birmingham*, Oxford, 1963, describes the work of the members. R. K. Merton, *Science, Technology, and Society in 17th-century England*, New York, 1970, a revision of an essay of 1938, is a classic study linking craftsmen with the rise of science.

8. J. Needham, *Science and Civilisation in China*, Cambridge, 1954, still in progress; P. J. P. Whitehead and P. I. Edwards, *Chinese Natural History Drawings selected from the Reeves Collection*, London, 1974; Sung Ying-hsing, *T'ien-kung k'ai-wu: Chinese Technology in the 17th Century*, tr. E-tu Zen Sun and Shiou-Chuan Sun, London, 1962. S. Nakayama, D. L. Swain, and E. Yagi, *Science and Society in Modern Japan: Selected Historical Sources*, Tokyo, 1974. Dharampal, *Indian Science and Technology in the 18th century*, Delhi, 1971; *Icones Roxburghianae, or Drawings of Indian Plants*, Calcutta, 1964, still in progress; M. Archer, *Natural History Drawings in the India Office Library*, London, 1962; A. Rahman *et al.*, *Bibliography of Source Material on History of Science and Technology in Medieval India*, New Delhi, 1976. D. J. de Solla Price, *Little Science, Big Science*, New York 1965, draws attention to the different scale of twentieth-century science; and J. R. Ravetz, *Scientific Knowledge and its Social Problems*, London, 1971, describes the problems that arise from the increased size and importance of science. On revolutions in recent science, see J. D. Watson's entertaining fragment of autobiography, *The Double Helix*, London 1968; and R. C. Olby's thorough work, *The Path to the Double Helix*, London, 1974; and on Continental Drift, A. Hallam, *A Revolution in the Earth Sciences*, Oxford, 1973.

Suggested Further Reading

This list is not meant to be exhaustive, or even to pay all my intellectual debts, but to give an idea of what is available for those who want to pursue some topic. In addition there are various standard works of reference: C. Singer, E. J. Holmyard, A. R. Hall, and T. I. Williams (ed.), *A History of Technology*, 5 vols., Oxford, 1954–58; T. I. Williams (ed.), *A Biographical Dictionary of Scientists*, London, 1969; A. Debus (ed.), *Who's Who in Science*, Chicago, 1968; and the multi-volume *Dictionary of Scientific Biography*, New York, 1970 still in progress. There are national societies for the history of science; that in the USA publishes the journal *Isis* containing papers, reviews and bibliographies, and that in Britain publishes the *British Journal for the History of Science*. The Society for the Study of Alchemy and Chemistry publishes *Ambix*, a journal devoted to the history of chemistry; and there are other specialised societies and journals. There are some journals independent of societies, such as *Annals of Science* which publishes original papers, and *History of Science* which contains review articles. There is a Society for the Bibliography of Natural History, which publishes its journal; and the British Museum (Natural History) publishes an historical *Bulletin* with papers on the history of the field. For the history of technology, there is the Newcomen Society and its *Transactions*. Dealing with exploration and naval history, there are the Navy Records Society and the Hakluyt Society, both of which publish very carefully edited and selected documents. The various societies also organise conferences from time to time; and at intervals there are international congresses on the history of science and technology.

INDEX

Index